Wilhelm Bauer, Wilfried Sihn, Peter Ohlhausen (eds.)

Journal of Engineering, Management and Operations

Volume I

Journal of Engineering, Management and Operations

edited by

Prof. Dr.-Ing. Prof. e.h. Wilhelm Bauer
(Fraunhofer IAO Stuttgart)

Univ.-Prof. Dr.-Ing. Dr. h.c. Dipl.-Wirtsch.-Ing. Wilfried Sihn
(Fraunhofer Austria Wien)

Prof. Dr. Peter Ohlhausen
(Fraunhofer IAO Stuttgart
und ESB Business School der Hochschule Reutlingen)

Volume 1

LIT

Wilhelm Bauer, Wilfried Sihn, Peter Ohlhausen (eds.)

Journal of Engineering, Management and Operations

Volume I

LIT

This book is printed on acid-free paper.

Bibliographic information published by the Deutsche Nationalbibliothek
The Deutsche Nationalbibliothek lists this publication in the Deutsche Nationalbibliografie; detailed bibliographic data are available on the Internet at http://dnb.d-nb.de.

ISBN 978-3-643-99768-5 (pb)
ISBN 978-3-643-99268-5 (PDF)

A catalogue record for this book is available from the British Library.

Klosbachstr. 107
CH-8032 Zürich
Tel. +41 (0) 44-251 75 05
E-Mail: zuerich@lit-verlag.ch   http://www.lit-verlag.ch
Distribution:
In the UK: Global Book Marketing, e-mail: mo@centralbooks.com
In North America: Independent Publishers Group, e-mail: orders@ipgbook.com
In Germany: LIT Verlag Fresnostr. 2, D-48159 Münster
Tel. +49 (0) 2 51-620 32 22, Fax +49 (0) 2 51-922 60 99, e-mail: vertrieb@lit-verlag.de
e-books are available at www.litwebshop.de

Vorwort

Exegi monumentum aere perennius
Horaz

Der erste Band des „Journal of Engineering, Management and Operations" ist fertiggestellt. Viele sagen: Noch ein Journal. Wir sind allerdings der Ansicht, dass es gerade in der Diskussion um Grundlagenforschung und der angewandten Forschung keine zielführende Entsprechung in der Publikationswelt gibt.

So haben wir uns als Fraunhofer-Institut für Arbeitswirtschaft und Organisation IAO in Kooperation mit der Universität Stuttgart und der Fraunhofer Austria Wien, die u.a. mit der Technischen Universität Wien verbunden ist, dazu entschlossen, diese Lücke zu schließen. Schwerpunkt der ersten Ausgabe ist das aktuelle Thema „Industrie 4.0". Der Begriff beinhaltet viele Ambivalenzen der aktuellen Diskussion in der Wissenschaft und der Industrie, ausreichend offen für die „Unvorhersehbarkeit der Zukunft". Mit dem Journal wollen wir einen Überblick über Fragestellungen von Unternehmen zu den drei Fokusthemen „Digitalisierung der Fabrik", „Logistik" und „der Mensch in der Fabrik der Zukunft" geben.

Für die Beantwortung der Fragen stellten sich verschiedene Vertreterinnen und Vertreter aus Unternehmen mit Ihrer Expertise zur Verfügung, die in Zusammenarbeit mit den Professoren der Hochschule Reutlingen Lösungen für ihre Betriebe erarbeiteten. Diese Erkenntnisse sind in die Beiträge intensiv eingearbeitet worden.

Besonderen Dank gilt den Kolleginnen und Kollegen Prof. Dr.-Ing. Vera Hummel für ihr Engagement zum Zustandekommen des ersten Bandes und Prof. Dr. Volker Reichenberger, der unermüdlich und mit nie enden wollender Kraft den ganzen Review-Prozess gesteuert und die Redaktion der Beiträge übernommen hat.

Dieses Journal ist das erste Werk in der Reihe. Innerhalb der Reihe werden wir zukünftig besondere Ereignisse und herausragende Projektergebnisse aus der angewandten Forschung präsentieren. Ein Dank geht hier auch an den Verlag, nicht nur durch die freundliche und unkomplizierte Zusammenarbeit, sondern auch für die Bereitschaft, langfristig mit uns zusammen zu arbeiten.

Die Herausgeber

Inhaltsverzeichnis

Contributions in English

Beiträge in Deutscher Sprache

Contributions in English

Tools, Methods and Standards for the Assessment of Material Efficiency Aspects in Small and Medium Sized Enterprises (SMEs)

Anja Braun[a], Peter Kleine-Möllhoff[a], Volker Reichenberger[a], Stephan Seiter[a]

[a] *ESB Business School, Reutlingen University, D-72762 Reutlingen*

Abstract

Mature economies which are driven mainly by small and medium sized enterprises (SMEs) are increasingly becoming dependent on material imports. Global material consumption is ever increasing, mainly driven by population increase. Decoupling of material consumption from economic growth is one of the greatest challenges of the 21[st]century. Within this paper available methods for the assessment of material efficiency on different economic scales are investigated and those detected that are particulary suitable for the use in SMEs. Recommendations for further improvements of the selected tools and an outlook concerning planned research activities in the field of material efficiency in enterprises, supply chains and circular economy aspects are given.

1 Introduction

Global population increased fourfold in the 20th century (Liedtke et al., 2014, p. 545). The global gross domestic product (GDP) and incomes grew stronger than the global material use since the 1950s, indicating an increase of resource productivity (Campanale and Femia, 2013, p. 614; Schaffartzik, Wiedenhofer, and Fischer-Kowalski, 2016, p. 201). Most of this productivity improvement can be attributed to the decline of biomass intensity in the last century (Steinberger, Krausmann, and Eisenmenger, 2010, p. 1156). The absolute levels of material use increased on a global scale and are mainly driven by population growth (Schaffartzik et al., 2016, p. 202). Between 1900 and 2010 global material consumption increased from approximately 6.6 to 70 Gt/a. Within the 50 years after 1960 the global material consumption almost quadrupled (Campanale and Femia, 2013; Giljum, Behrens, Hinterberger, Lutz, and Meyer, 2008; Hoang, 2014; Liedtke et al., 2014; Schaffartzik et al., 2016; Schandl et al., 2016). Since 2000, the global material productivity stagnates (Campanale and Femia, 2013).

Mature economies like the European Community (EU) are increasingly becoming dependent on material imports, especially concerning fossil fuels and metal ores. Europe covers 83% of iron ores, 80% of bauxite and 74% of copper demand by imports (Giljum et al., 2008).

Under life cycle impact considerations, mass materials such as iron ore, aluminium and steel are as environmentally intensive as fossil fuels (Bleischwitz, Welfens, and Zhang, 2012, p. 56). Steel production accounted e.g. for 25 % of global industrial CO_2 emissions (Wiedmann, Schandl, and Moran, 2015, p. 371).

Small and medium sized enterprises (SME) represent worldwide the economic backbone (OECD, 2017). Within the EU28 manufacturing stood in 2015 for 44 % of total value added by SMEs (European Commission, 2016). Concerning environmental friendly activities however, SMEs are perceived as underperforming (Lewis and Cassells, 2010, p. 8). Material efficiency improvement efforts are often considered to conflict with economic objectives and appear to involve complex analysis and implementation actions. If material losses can be compensated by profits, it seems that they are ignored as a management goal (Kokubu and Kitada, 2015, p. 1281).

A large amount of tools and methods are available to support the reduction of environmental burdens in manufacturing. The purpose of this paper is to analyse and compile the current knowledge in the area of material efficiency improvement tools, guidelines and standards and to detect those which may support most effective efforts of SMEs to reduce their material intensity in manufacturing.

2 Research methodology

For the investigations of available tools, methods and standards concerning the improvement of material efficiency an extended literature review for the time between 2006 and January 2017 was conducted aiming to collect and analyze all relevant papers in the field of interest by using a structured search of literature. These have been sourced from Academic Search Premier, Business Source Complete, China Online Journals, Complementary Index, Directory of Open Access Journals, EBSCO, Emerald Insight, IEEE Xplore Digital Library, JSTOR, OaFindr, Science Direct, SpringerLink, Supplemental Index, Taylor and Francis, TEMA and Wiley. Appropriate keyword combinations have been used ranging from 'assessment', 'benchmark', 'cleaner production', 'consumption', 'dematerialization', 'efficiency', 'ecological footprint', 'ecosystem', 'environmental accounting', 'flow cost accounting', 'industrial ecology', 'mapping', 'modeling', 'framework', 'intensity', 'manufacturing', 'material', 'production', 'productivity', 'raw material consumption', 'resource', 'reduction', 'sustainability', 'waste flow'. The academic references have been scanned and selected concerning the relation of their title and abstract content to the investigation field and rated by the authors. In addition to the papers selected for complete review, publications covering material efficiency and environmental concerns from the International Organization for Standardization (ISO) and the German Engineering Associations (VDI) standards have been investigated as well. The latter are avaiable bilingual in German and English language. Citations have been examined, those not retrievable and not covering well enough the related sense have been discarded.

The retrieved tools and methods have been categorized concerning their superordinate concept and analyzed concerning their comprehensiveness, the assessed metrics, the economoic scale of usability and their entrance barriers for SMEs.

Finally a selection of suitable tools and methods is taken, the results are discussed and the reqirements for improvement for SME use are described.

3 Analysis of tools, methods and standards supporting material efficiency

Zhong et al. (2016) reviewed the research progress especially on natural resource accounting between 1995 and 2017 and found that water and energy are the resources which found most attraction in the last 20 years (Zhong et al., 2016, p. 122). Mathematic modeling, emergy, exergy, environmental footprint (EF), life cycle assessement (LCA) and material flow analysis (MFA) seem to be common methods to account those natural resources and can be classified into economic and eco-centric methods (Zhong et al., 2016, pp. 129-131). As none of the single methods can solve all the problems, hybrid approaches are often considered or recommended (Nakajima and Ortega, 2016; Wang et al., 2015). Chomkhamsri and Pelletier (2011) analyzed existing EF and LCA methodologies for products and organizations used in the European Community in order to form a starting point for the development of a harmonized European methodology for EF that can accommodate a broad suite of relevant environmental performance criteria. The analysis does not include the Material flow cost accounting (MFCA) approach. However, it asserts that existing EF and LCA methods and standards are a challenge for the company side as life cycle approaches have been used mainly for product evaluation and have not played an important role on a company assessment level (Chomkhamsri and Pelletier, 2011, p. 3).

Angelakoglou and Gaidajis (2015) reviewed existing methods contributing to the triple bottom line (TBL) assessment of industrial systems and classified those (Angelakoglou and Gaidajis, 2015, pp. 738-744). Ten of the reviewed methods are suitable for material and energy efficiency purposes and fall into the classes energy and material flow analysis (EMFA), environmental accounting (EA) and LCA.

Beside those most recent reviews and analysis of methods that fall in the broader sense of EA and are suitable for material efficiency purposes, a cornucopia of tools and methods is available in the investigated literature. Although there are ongoing terminology-related standardization efforts, many terms used in the area of sustainable manufacturing are not well defined, undetermined or overlap in meaning. A common set of terms and definitions is to be developed (Garretson, Mani, Leong, Lyons, and Haapala, 2016, p. 987). Therefore the retrieved tools and methods have been categorized and defined as shown in Table 1 concerning their superordinate concept. In total, ten superordinate categories have been defined and their related methods, tools and standards are briefly described and assessed in the following sections in detail. EA has not been chosen as superordinate concept as it is too general and not

Concept	Abbreviation	Concept	Abbreviation
Cleaner production	CP	Project Management	PM
Eco-Efficiency	EE	Change Management	CM
Energy and Material Flow Analysis	EMFA	Exergy and Emergy	EXEM
Life Cycle Assessment	LCA	Triple Bottom Line Approach	TBL
Lean Management	LM	Miscellaneous concepts	
Material Flow Cost Accounting	MFCA		

Table 1: Superordinate concepts and abbreviations

	Metric						Scale				
Method/Tool/ Standard	**Energy**	**Material**	**Waste**	**Emissions**	**Costs**	**Social**	**Macro**	**Meso**	**Micro**	**Entrance barrier for SMEs**	**Latest publications**
Quick green scan	+	+	+	+	+	+			+	high	Kluczek (2017)
Ontological framework				+					+	high	Muñoz, Capón-García et al. (2013)
TRIZ	+	+	+	+	+			+	+	high	Kubota and da Rosa (2013)
Environmental management accounting	+	+	+	+	+				+	medium	Kurdve et al. (2015)
VDI 4075 part 1 Cleaner Production	+	+	+	+	+				+	medium	VDI Society Energy and Environment (2014)

Table 2: Cleaner production related tools and methods

specific enough. Those concepts which are not assignable to these categories are embraced under miscellaneous concepts.

3.1 Cleaner Production

Cleaner Production (CP) strives to increase the efficient use of energy and materials in production in order to reduce costs and environmental risks by applying an integrated strategy (Schaltegger, 2008, p. 8).

Table 2 illustrates the main properties of the tools retrieved and investigated concerning the cleaner production category. The table and those of the following sections of this chapter include only the latest publications. Earlier works to which the latest authors refer are named in the descriptive text for the tables. Kurdve et al. (2015) for instance, refer to earlier works from Jasch (2003) and Schaltegger (2008). All four CP-related tools concentrate on the micro-economic scale which means on a company boundary scale. The application of the theory of inventive problem solving (TRIZ) however, used by Kubota and da Rosa (2013) for CP-aspects includes a whole industrial branch (dairy industry in Brazil) and thus has been applied on a meso-economic scale. TRIZ has been identified as a strong tool especially when the availability of company data is scarce (Kubota and da Rosa, 2013, p. 199). In contrast to the other tools, TRIZ is not input, nor output related.

The quick green scan-tool from Kluczek (2017) is based on a fuzzy method especiallly designed for decision making in heating device manufacturing and covers also social metrics which makes it suitable for a TBL approach as well. The tool represents a high entrance barrier for SMEs, due to the fact requiring external expertise concerning LCA knowledge and its complexity.

The ontological framework presented by Munos (2013) delivers only emission (output) related environmental information. Further it is a complex conceptual approach which requires multi external know how and thus, represents a high entrance barrier for SMEs.

Environmental management accounting (EMA) delivers all required physical and monetary data on a company scale. However, it lacks a standard definition of environmental costs and represents a medium entrance barrier for SMEs.

Standards and official guidelines are also represented in the tables of this section and the following ones. In CP, the VDI 4075 part 1 (2014) is established since 2005. This standard is intended for SME practitioners in production wishing to reduce material consumption, emissions and costs of production processes and plants by using an integrated approach. Beside social metrics it delivers all relevant physical and economic metrics, concentrates on a company level and has been used at least in Germany longterm and in multiple industry branches. It requires external support and represents therefore a medium entrance level for SMEs.

3.2 Eco-Efficiency

Eco-efficiency (EE) is the combination of ecological and economic efficiency aspects by increasing resource productivity and reducing waste and pollution creation (DeSimone and Popoff, 1998, pp. 2-3). In Table 3 the EE-related tool properties are indicated.

The EMA tool, presented in Table 2 and described in 3.1 can be counted also as EE-tool. EE procedures are standardized by the DIN EN ISO 14045 (2012). However, this standard is specifically product related and thus, it is a management tool for environmental impacts of products and their value. The requirements of the standard represent a high entrance barrier for SMEs.

Cezar Lucato et al. (2013) present a conceptual mathematical approach to indicate the eco-efficiency of a single manufacturing process unit and then to extending indication to the enterprise (micro) level. The approach is clearly indication related, issues

	Metric						Scale				
Method/Tool/ Standard	**Energy**	**Material**	**Waste**	**Emissions**	**Costs**	**Social**	**Macro**	**Meso**	**Micro**	**Entrance barrier for SMEs**	**Latest publications**
ISO 14045 Eco-Efficiency	+	+	+	+	+				+	high	DIN-Normenausschuss Grundlagen des Umweltschutzes (NAGUS), Principles of Environmental Protection Standards Committee (2012)
Eco-Efficiency	+	+	+		+				+	medium	Cezar Lucato et al. (2013)

Table 3: Eco-efficiency related tools

are named and need to be resolved, further development is required before it can be used by SMEs.

3.3 Energy and Material Flow Analysis

Two analysed approaches can be subsummized under energy and material flow analysis (EMFA) aspects (see Table 4).

The waste flow mapping (WFM) tool presented by Kurdve et al. (2015) is a concept for waste management in supply chains (SC) (meso scale) and is based on lean managements value stream map (VSM). It includes energy flows, their related material flows and recycling aspects. The concept needs to be further developed especially concerning stakeholder management before it is usable in practice and for SMEs.

In contrast the material, energy and waste process flow modelling concept presented by Smith and Ball (2012) is the only only concept of all investigated tools that is tailored especially for the needs of SMEs and targeted for low complexity and thus a low entrance barrier. Top down and bottom up approaches are combined to assess material, energy and waste flows and their costs and to understand in- and outputs on all manufacturing system levels. In contrast to other tools (except the VDI 4075 part 1, 2014), techniques are presented to gather and assess the required data in an efficient way. The concept has been applied successfully with scientific external support in one industrial case and needs to be further tested and improved (Smith and Ball, 2012, pp. 229-237).

Nakamura et al. (2011) present a concept called UPIOM to overcome the issue of transferring sector and nationwide monetary input-output tables (MIOT) for socio-economic systems into physical input-output tables (PIOT). The highly scientific and complex approach concentrates on products, uses the Pareto principle and delivers a triangular visualization of material flows from macro to micro economic scale.

Method/Tool/ Standard	Metric						Scale			Entrance barrier for SMEs	Latest publications
	Energy	Material	Waste	Emissions	Costs	Social	Macro	Meso	Micro		
Waste Flow Mapping	+	+	+					+		n.a.	Kurdve et al., 2015
Material, Energy and Waste process flow modelling	+	+	+		+				+	low	Smith and Ball, 2012, p. 227
Unit Physical Input-Output by Materials (UPIOM)		+			+		+	+	+	high	Nakamura, Kondo, Matsubae, Nakajima, and Nagasaka, 2011

Table 4: Energy and material flow analysis related tools

3.4 Life Cycle Assessment related tools

Life cycle assessement (LCA) methods capture environmental impacts of processes and services from raw material acquisition to final discarding to the environment (DIN-Normenausschuss Grundlagen des Umweltschutzes (NAGUS), Principles of Environmental Protection Standards Committee, 2009, p. 4). The method is standardized and has reached a decent level of maturity. It is clearly oriented towards environmental impacts, thus emissions and resource depletion. The standards that are LCA related are displayed in Table 5.

The DIN EN ISO 14040 (2009) and 14044 (2006) are product and service related, take a cradle to grave perspective, require strictly a scientific approach and complete data acquisition. Know how to handle the databasis and software tools offered for LCA are unevitable. SMEs need to build up capacities in order to use LCA or to use external expertise, both reflects in most cases a high entrance barrier for such enterprises.

The VDI 4600 (2012) is a guideline for the determination of the total energy demand (KEA) of products and services from cradle to grave. The result of KEA describes the sum of cumulative non-renewable demand (KNRA) and the cumulative renewable demand (KRA).

The VDI 4800 part 1 (2016) and 2 (draft 2016) describe how to determine the cumulated raw material demand (KRA) which includes also raw material demand for energy. Secondary raw material (waste or residue which can be used as raw material, e.g. copper scrap) and which can replace primary raw materials are considered. The approach is strongly linked to the LCA methodology and concentrates mainly on the criticality of raw materials.

All LCA related methods and tools are listed in Table 6. Methods that are based on a similar approach, like e.g. the environmental footprint (EF) are also listed under the superordinate LCA-concept. There are in total fourteen retrieved methods and tools

	Metric						Scale				
Method/Tool/ Standard	Energy	Material	Waste	Emissions	Costs	Social	Macro	Meso	Micro	Entrance barrier for SMEs	Latest publications
ISO 14040, 14044 Life Cycle Assessment	+			+				+	+	high	DIN-Normenausschuss Grundlagen des Umweltschutzes (NAGUS), Principles of Environmental Protection Standards Committee (2009) DIN-Normenausschuss Grundlagen des Umweltschutzes (NAGUS), Principles of Environmental Protection Standards Committee (2006)
VDI 4600 Cumulative Energy Demand	+							+	+	high	VDI Society Energy and Environment (2012)
VDI 4800 part 1 and 2 Resource Efficiency		+						+	+	high	VDI Society Energy and Environment (2016a), (2016b)

Table 5: LCA related standards

that are related to LCA and had been investigated in detail. Many proposed concepts listed are hybrid approaches. The ISO 14040ff standards and many related tools do not focus on material and waste as long as there is no environmental impact, which limits their use for manufacturing SMEs. Five concepts cover a macroeconomic scale. These are shown at the end of Table 6 and are not further discussed here as a micro and mesoeconomic view is important for SMEs in first instance.

Latham (2016) combines LCA with an economic assessment for remanufactured products (light trucks). Herva et al. (2012) and Herva et al. (2008) propose a spreadsheet based approach for the EF related methods in order to facilitate data collection. Saravia-Cortez, Herva et al. (2013) present a case from wood-board manufacturing in Spain, where EF methodology has been used. This method indicates land use which does not support decision making on a micro economic scale and thus is not suitable for practitioners in SMEs.

The Gaia Biorefinertool proposed by Rönnlund et al. (2016) is a special solution for the metals industry and requires a specific software offered by a metals furnace supplier in Finland.

Boonsothonsatit et al. (2015) propose a generic decision support system based on multi-objective optimisation for green SC network design (GOOG). A multi-objective function is converted into a fuzzy membership function and finally converted into a Pareto-optimal function.

The environmental impact of materials and transportation is calculated based on the streamlined-LCA (SLCA) database of Kara, Manmek, and Kaebernick, 2007. The

Method/Tool/ Standard	Metric						Scale			Entrance barrier for SMEs	Latest publications
	Energy	**Material**	**Waste**	**Emissions**	**Costs**	**Social**	**Macro**	**Meso**	**Micro**		
Economic Input-Output Life Cycle Assessment	+			+	+			+	+	high	Latham (2016)
Energy and Material Flow Management + EF	+	+	+	+				+	+	medium - high	Herva et al. (2012)
Environmental Footprint (EF)	+		+	+					+	high	Saravia-Cortez et al. (2013) Herva et al. (2008)
Gaia Biorefiner enhanced	+			+				+	+	high	Rönnlund et al. (2016, p. 1473)
Generic decision support system based on multi-Objective Optimisation for Green SC network design (GOOG)	+			+	+			+		high	Boonsothonsatit et al. (2015)
LCA for multirecycling	+			+				+	+	high	Neugebauer et al. (2013)
Material Input per Service unit (MIPS)		+							+	high	Saurat and Ritthoff (2013, p. 582), Liedtke et al. (2014, p. 545), Geibler et al. (2016, p. 72)
Streamlined LCA	+			+					+	high	Graedel, 1998; Jayal et al. (2010, p. 145)
Dynamic Material Flow Modelling				+			+			high	Giurco and Petrie (2007, p. 842)
Economic Input Output + Data Envelopment Analysis	+			+	+		+			high	Egilmez et al. (2016, p. 622)
Material Flow analysis + Economic Input/Output modelling		+	+	+	+		+			high	Hawkins et al. (2007, p. 1024)
Material Flow Analysis + Structural Agent Analysis		+	+	+			+			high	Binder (2007, p. 1605)
Material Flow analysis + Value Chain Analysis		+	+	+	+		+			high	Dahlström and Ekins (2007, p. 542)

Table 6: LCA related tools and methods

approach is not designed for the application on a micro level.

In order to evaluate proper environmental burdens of materials which can be recycled without quality loss (e.g. steel), Neugebauer et al., 2013, 52ff abandon the product approach of LCA in favor of a material pool approach. The environmental burdens are added up over several life cycles. This means that the total environmental burden reduces with increasing life cycles. The approach is rather of interest for industry branches than for SMEs.

Egilmez et al. (2016) present a combination of LCA and economic input output (EIO) analysis for the macroeconomic evaluation of the food manufacturing industry, using a fuzzy data envelopment analysis (DEA) approach in order to consider potential data uncertainties. The environmental footprint (EF) approach proposed by Herva et al. (2008) and Saravia-Cortez et al. (2013) is applicable for products and processes. The EF final result is one single indicator concerning the land use of all operations involved in manufacturing of a product or in operations and is based on many assumptions concerning background processes and the provision of energy. Thus it does not deliver clear indications for SME decision makers. This is a general disadvantage of LCA-approaches in the context of manufacturing process assessment and optimization. LCA-type approaches provide plenty of environmental impact information for hotspots, they do neither provide a solution nor a techno-economic basis to improve the system (Reuter, van Schaik, and Gediga, 2015, p. 683).

The material input per service unit (MIPS) approach has been developed in the early 1990s reaching since then a decent maturity on one side and a certain complexity on the other side. As in EF, one indicator for the cumulative primary materials demand needed over the whole life cycle of a product relative to its service will be the final indication (Saurat and Ritthoff, 2013, p. 582). Unused extraction which does not appear in consumption statistics is included in the MIPS approach (Liedtke et al., 2014, p. 548). These are however, based on assumptions concerning background systems. A connection to processes and economic metrics is feasible. Geibler et al. (2016) present a mixed-method approach by using a combination of two screening LCA methods and showed in a practical example that the combination of MIPS and hot spot analysis (HSA) are less time consuming than a full LCA (Geibler et al., 2016, p. 72). The proposed method is only feasible with an expert in MIPS and a significant amount of data has to be collected from literature. Further the material input factors do not show which raw materials specifically are used, only summarized biotic, abiotic resources are compiled in the concept.

Jayal et al. (2010) reference a publication from Graedel (1998) concerning streamlined LCA in order to reduce the efforts for application. The implementation hurdle is still high and they concentrate like Boonsothonsatit et al. (2015) on the optimization of SCs.

The majority of LCA related concepts represent a high implementation hurdle for SMEs due to their complexity, data requirements and scientific approach. Only Herva et al. (2012) and Herva et al. (2008) try to reduce the implementation hurdle by using a spreadsheet based approach.

	Metric						Scale				
Method/Tool/ Standard	**Energy**	**Material**	**Waste**	**Emissions**	**Costs**	**Social**	**Macro**	**Meso**	**Micro**	**Entrance barrier for SMEs**	**Latest publications**
Lean Management + LCA				+					+	high	Erlach et al. (2012)
VSM + MFCA + Pinch method		+	+		+				+	medium	Thanki and Thakkar (2016)
Environmental VSM	+		+	+	+				+	unknown	Kurdve et al. (2015)

Table 7: Lean management related tools and methods

3.5 Lean Management

One concept that uses lean managament (LM) approaches, the WFM has been described already in 3.3. Three more concepts based on LM have been identified and are listed in Table 7. The LM + LCA approach from Erlach et al. (2012) requires data acquisition based on real measurements which get implemented into LCA datasets. Based on a value stream analysis known from LM, some additional ecological data are collected. The method delivers CO_2 equivalent indicators and does not support directly material and waste reduction efforts.

Thanki and Thakkar (2016) propose to combine VSM with MFCA and the pinch method. The pinch method detects favorable thermodynamic potentials and quantity differences, in order to use these potentials most efficient in a cascade on different levels. The thermodynamic pinch concept can be transferred as well to material and waste costs. The total manufacturing costs are compared with the value according to lean management thinking in order to target areas of improvement in manufacturing. The empirical scientific method concentrates mainly on idle times in manufacturing. It depends on a computer based support system and needs to be further developed concerning the integration of information flows, overproduction, inventory and emission issues (Thanki and Thakkar, 2016, pp. 1285-1295).

Kurdve et al. (2015, p. 306) describe beside the WFM, presented in 3.3 in addition an environmental VSM (E-VSM) to map material and energy use in enterprise processes along with the activities, time and information in the process including lead times and inventory. No concrete information is given about the use of this concept in practice. Kurdve et al. (2015, p. 313) admit that LM approaches focus in general on waste of time rather than on material efficiency.

	Metric						Scale				
Method/Tool/ Standard	**Energy**	**Material**	**Waste**	**Emissions**	**Costs**	**Social**	**Macro**	**Meso**	**Micro**	**Entrance barrier for SMEs**	**Latest publications**
ISO 14051 Material Flow Cost Accounting	+	+	+		+			+	+	high	DIN-Normenausschuss Grundlagen des Umweltschutzes (NAGUS), Principles of Environmental Protection Standards Committee (2011)
MFCA + LCA	+	+	+	+	+			+	+	high	Nakano and Hirao (2011, pp. 1190–1192)
MFCA + Microsoft Excel	+	+	+		+				+	high	Kokubu and Kitada (2015)
MFCA + dynamic investment calculation	+	+	+		+				+	high	Schmidt et al. (2015, p. 1331)
MFCA + LCA + LCC	+	+	+	+	+			+	+	high	Bierer et al. (2015)
MFCA/WIP software simulation	+	+	+		+				+	high	Zhao (2012)
All Seeing Eye of Business			+		+			+	+	medium	Bautista-Lazo and Short (2013)
Optimized MFCA for circular economy aspects	+	+	+		+			+		high	Zhouet al. (2017)
Extended Activity-based environmental costing		+	+		+				+	high	Cagno, Micheli, and Trucco (2012)

Table 8: Material Flow Cost Accounting related tools and methods

3.6 Material Flow Cost Accounting

Material Flow Cost Accounting (MFCA) according to the ISO 14051 standard is a management tool that can assist organizations to achieve both environmental and financial improvements via changes in their material and energy use practices those practices (DIN-Normenausschuss Grundlagen des Umweltschutzes (NAGUS), Principles of Environmental Protection Standards Committee, 2011, p. 9). Beside that standard, the properties of related tools are listed in Table 8. The standard is usable on a company level and in SCs and requires, as all related tools listed, a certain scientific level and completeness which represents per se mainly a high entrance barrier for SMEs. MFCA is like LCA often opaque, costly, and time consuming (Garretson et al., 2016, p. 987). Sygulla, Götze, and Bierer (2014, 115ff) recommend to improve MFCA concerning the appropriate treatment of inventories, storage costs, the flow modelling of energy the evaluation of future states in process chains. They criticize the MFCA's limitation to a single time period and propose to complement MFCA with dynamic investment appraisals. Bierer et al. (2015) try to overcome some of these issues by combining MFCA with LCA and Life Cacle Costing (LCC). They highlight however, to evaluate carefully the balance between the necessary efforts of such an approach and the possible benefits and propose therefore to concentrate in a first step only on important materials. The combination of MFCA with dynamic investment calculation presented by Schmidt, Götze, and Sygulla (2015, p. 1331) is a usable concept. It contains cashflow oriented elements, which should be more attractive for decision makers than static historically oriented economic assessment. However, the concept should be further developed for SME use.

Zhao (2012) combined MFCA with the work in progress (WIP) approach in a software simulation provided by Rockwell Automation for a case study. A centralized fuzzy control methodology is used to control the change in WIP inventory level and to assess the related costs. The approach is in conceptual stage.

The all seeing eye of business concept, presented by Bautista-Lazo and Short (2013) represents a medium entrance barrier, covers however, only waste aspects and is product design oriented. Cagno et al. (2012) propose to extend a general activity-based environmental costing approach by an MFCA approach which includes the identification of activity drivers and quantification of their intensities. The tool has been tested in an SME which had highly motivated emloyees and still took two months for its implementation. The method requires a stong effort and is quite time-consuming and can be very expensive unless a management system already exists (Cagno et al., 2012, p. 142). The combination of MFCA and LCA proposed by Nakano and Hirao (2011, pp. 1190–1192) is designed for the optimization of SCs. Zhou et al. (2017) present a complex concept which concentrates on recycling aspects. Kokubu and Kitada (2015, p. 1287) have shown in several case studies in Japan that for an effective use of MFCA approaches modification of organizational design in terms of the span of control and accountability has to be made in order to enable mutual adjustment.

3.7 Project management based tools

The span of control and accountability mentioned in section 3.6 could be achieved if material efficiency improvement measures would be handled as projects. Such an approach would facilitate the implementation for SMEs. Four project management (PM) related approaches for material efficiency purposes have been found and investigated (see Table 9), all of them represent a medium barrier for SMEs.

The VDI 4090 standard is a project management standard that delivers guidance to implement environmental related projects in enterprises but does not contribute to any metrics.

The concept presented by van Berkel (2007) and the methodology of Schliephake et al. (2009) are SC or even recycling oriented. Only the latter covers explicit physical metrics, the other two monetary and management related parameter. The DMAIC

	Metric						Scale				
Method/Tool/ Standard	**Energy**	**Material**	**Waste**	**Emissions**	**Costs**	**Social**	**Macro**	**Meso**	**Micro**	**Entrance barrier for SMEs**	**Latest publications**
VDI 4090 Systems engineering methodology for design and control of environmental relevant processes in the operational area								+	+	medium	VDI Society Energy and Environment (2005)
Define, Measure, Analyse, Improve and Control (DMAIC) methodology	+	+	+					+		medium	Schliephake, Stevens, and Clay (2009, p. 1258)
McKinsey 7S framework				+	+				+	medium	Teh and Corbitt (2015)
Eco-Efficiency implementation routine					+			+		medium	van Berkel (2007, 533ff)

Table 9: Project management related tools and methods

methodology delivers a step by step procedure which is favorable for SMEs. The 7S Strategy development concept presented by Teh and Corbitt (2015) represents a management tool which combines product and strategic dimensions for value creation.

3.8 Change management related tools

Material efficiency improvements mostly require organizational modifications. Material or energy losses are often caused by organizational inefficiencies. Attacking those inefficiencies by organizational or behavior changes is the first and often the economic most favorable choice (immediate return on investment). If the inefficiencies have technical reasons, the modification or replacement of existant technology often also entails organizational changes. Therefore, change management (CM) is an important aspect which requires more attention in practice.

One publication has been retrieved that covers this important aspect. Sekerka and Stimel (2011) propose a CM strategy development concept based on value reflection and relationship management. However, the presented practical concept for eco-sustainability is strongly related to the situation in recession times after 2008. Therefore in order to be useable as a general CM tool for SMEs further developments are necessary.

	Metric						Scale				
Method/Tool/ Standard	**Energy**	**Material**	**Waste**	**Emissions**	**Costs**	**Social**	**Macro**	**Meso**	**Micro**	**Entrance barrier for SMEs**	**Latest publications**
Exergy	+						+		+	medium	Ignatenko et al., 2007, p. 863, Hoang, 2014, Simon and Holm, 2017, Miller et al., 2016, Yang et al., 2015, Szargut, 2005
Emergy	+	+							+	high	Hoang, 2014, p. 130

Table 10: Exergy and emergy related tools and methods

3.9 Exergy and emergy related tools

Exergy represents the maximum quantity of potential work that a system can execute until it is in equilibrium with its reference environment (Wall, 1977, p. 4). Ignatenko, van Schaik, and Reuter (2007) and Simon and Holm (2017) used exergy especially for the evaluation of recycling systems, Miller, Foxon, and Sorrell (2016) and Yang, Yang, and Qian (2015) used the exergy approach for nationwide macro economic investigations (see Table 10). The latter has been used to quantify the monetary value of natural services. The exergy is in principle a useful physical metric to quantify the quality of energy and materials (which contain energy). However, the methodology has not been applied consistently up to date.

For SME decision making it is suitable to assess and to decide for different recycling options. For a general use concerning material efficiency aspects however, it is not suitable for the SME field. Further, when enterprises manage to operate manufacturing systems in the future mainly based on renewable energy sources and secondary (recycled) materials, an exergy analysis becomes obsolete.

Emergy evaluation traces solar energy embodied in a product and thus, can be used to describe flows of matter and energy (Hoang, 2014, p. 130). Elevated knowledge is required to use this approach and as a product related method it is less usefull for decision making in manufacturing.

3.10 Triple Bottom Line based approaches

Five methods have been retrieved that include beside physical and monetary also social aspects thus, the triple the bottom line (TBL) approach (see Table 11).

Turkay, Saracoglu, and Arslan (2016, p.1) developed a method for enterprises that adds criteria of a TBL approach to traditional cost accounting. The conceptual model has been tested in one real manufacturing case in Turkey.

The combination of Ultra-f-check and VSM shall enable companies to push their production systems towards greater efficiency and effectiveness (Kuhlmann and

	Metric						Scale				
Method/Tool/ Standard	Energy	Material	Waste	Emissions	Costs	Social	Macro	Meso	Micro	Entrance barrier for SMEs	Latest publications
Traditional cost accounting enhanced	+				+	+			+	medium	Turkay et al. (2016, p. 1)
Ultra-f-check + value stream method (VSM)	+	+	+	+	+	+		+	+	high	Kuhlmann and Sauer (2016, p. 482), Erlach et al. (2012)
Sustainability index modelling	+		+	+	+	+			+	medium	Latif et al. (2017)
Sustainable supplier evaluation				+	+	+		+		unknown	Kannanet al. (2013)
Metaheuristic modelling				+	+	+		+	+	high	Devika, Jafarian, and Nourbakhsh (2014)

Table 11: Triple bottom line related methods and tools

Sauer, 2016, p. 482). The method provides a quantitative mapping and a classification of maturity level for the elements of the factory in each action field (Kuhlmann and Sauer, 2016, p. 484). In total, hundred measurement values and key performance indicators (KPI's) have to be updated and users depend on an external database. In a reduced manner, capturing only most important values for material efficiency, the concept would represent a usable basis for SMEs especially as it includes recycling aspects.

Latif et al. (2017) propose an interactive questionnaire based model, covering three TBL indicators to assess companies. Performance indicator is the ratio of the positive responses to the total number of responses and is represented in a normalized form. Users also depend on an external database. The idea is suitable to create dynamic improvement by implementing continuously best practices into the questionnaire.

Kannan et al. (2013) propose a multi criteria decision methods based on fuzzy logic for supplier evaluation in order to reduce negative environmental and social impacts of company activities. The concept is designed to decrease the subjectivity of decision-makers' preferences, has been illustrated by a numerical example (Alma, 2013, p. 6) and needs to be further investigated for practical use.

Devika et al. (2014) propose and compare novel metaheursitc algorithms in order to simultaneously consider the three pillars of sustainability in SC networks. To evaluate the usefulness and performance of the proposed model and NIV. The method is in experimental status and needs further development for use by SMEs.

	Metric						Scale				
Method/Tool/ Standard	**Energy**	**Material**	**Waste**	**Emissions**	**Costs**	**Social**	**Macro**	**Meso**	**Micro**	**Entrance barrier for SMEs**	**Latest publications**
Cost Benefit Analysis					+				+	medium	Loiseau et al. (2016, p. 366)
Green-component life-cycle value design		+			+				+	unknown	Chung and Wee (2011, p. 202)
Aggregated process-level material input-output models	+	+	+	+					+	high	Xue, Kumar, and Sutherland (2007, p. 1349)
Particle Swarm Optimization	+	+			+				+	unknown	Hassine, Barkallah, and Bellacicco (2015)

Table 12: Miscellaneous methods and tools

3.11 Miscellaneous methods and tools

Table 12 shows tools and methods which have been retrieved during the research and cannot be allocated to the superordinate concepts shown in Table 1.

Loiseau et al. (2016) provide an overview concerning weak and strong sustainability tools for a product related green economy and propose to combine a discounted cost-benefit analyis (CBA) and total economic value (TEV) for the value of future streams with instruments like LCA, LCC, input-output-modeling (IO), environmentally extended input-output modeling (EEIO) (Loiseau et al., 2016, pp. 364–367).

Chung and Wee (2011) consider circular economy (CE) aspects and propose a model for a reverse supply chain for remanufactured electronic products. The investigations include the total costs reduction of the reverse supply chain while minimizing environmental burdens.

Xue et al. (2007) show with their aggregated process-level material input-output model how manufacturing processes can be mathematically modelled and how an input-output mass matrix can be solved, if there are several outputs to single inputs. Two cases from the automotive and steel industry have been used to shown how this approach can be used to minimize waste formation and to reduce emissions.

Hassine et al. (2015) show how particle swarm optimization (PSO) can be used in order to find the optimum machine setting in manufacturing concerning production time, costs and energy demand. The concept is very specific for the optimization of turning machines and needs to be further developed in order to be usable for other machining and manufacturing processes.

4 Summary and outlook

Most tools and methods retrieved in the literature research and compiled in this paper represent a medium or high implementation barrier for SMEs. Most tools are too involved, focused on and specialized for singular aspects, need specific external know-how for implementation or represent a conceptual state of the art.

Most environmental sustainability assessment methods fail to serve as a decision making tool, basically due to the absence of a specific sustainable reference point, their complexity and difficulty to be applied by non-experts (Angelakoglou and Gaidajis, 2015, 736ff). Standardized methods like LCA and MFCA are often opaque, costly, and time consuming (Garretson et al., 2016, p. 987) or even limiting when alternatives have to be assessed (Steubing, Mutel, Suter, and Hellweg, 2016, p. 520). There is clearly an unbalance between necessary efforts and possible benefits. The lack of business perspective, being too academic, difficult to perform, data availability (Allen et al., 2002, p. 913) and neither providing a solution nor a techno-economic basis to improve systems (Reuter et al., 2015, p. 683) are further hurdles for SMEs. The indicators provided, are often of scientific nature and on a macro economic scale (e.g. cumulated land use in EF) and thus not suitable for decision making on a micro economic scale. Further, most tools are limited to a single time period. Many proposals are on a conceptual stage and need further development before they can be applied in practice.

Only the EMFA tool presented by Smith and Ball (2012) and the CP guideline of the VDI Society Energy and Environment, 2014 are tailored for the process and enterprise gate to gate level and deal with material, waste and energy. The first can be rated to constitute a low, the latter a medium entrance barrier for SMEs as it requires external support for implementation. Both are tailored especially for the needs of SMEs and targeted for low complexity. Smith and Ball (2012, p. 229) recommend explicitly to combine a top down and bottom up approach in order to assess material, energy and waste flows and their costs and to understand in- and outputs on all manufacturing system levels. The concept has been applied successfully with scientific external support in one industrial case and needs to be further tested and improved in different areas (Smith and Ball, 2012, p. 237).

Material efficiency concerns technical, organizational and relationship matters on the enterprise and the SC levels and calls for project management approaches in order to achieve control and accountability. In order to encourage SMEs to start implementing material efficiency activities, acceptable instruments must bear a maximum of simplicity and transparency. With a limited amount of data to be acquired or provided a maximum short term economical benefit must be realized in order to convince SME decision makers. SMEs are not able or are not willing to supply comprehensive internal data. Therefore, a suitable tool for SMEs must abandon the concept of complete data acquisition and must work most reliable with incomplete data sets in order to collect low hangig fruit.

The path to implement material efficiency approaches should be as clear as possible. External support should be avoided or limited to the lowest level. Existing internal manpower must be able to handle such projects with little extra effort and must be confident to use material efficiency improvement instruments with minimal training demand. For decision making, dynamic financial appraisals are important. External change agents, introducing and promoting adoption of material efficiency by SMEs should be considered. However, such agents would need to buy-in the opinion leaders first (Sulong, Sulaiman, and Norhayati, 2015, p. 1369). This aspect calls for governmental or institutional support.

Once, SMEs have implemented material efficiency approaches it is important to collect information about process related topics in order to further develop a simplified method. It is important to report success stories but it is important as well to get a better understanding of practical implementation issues. This aspect calls for a scientific monitoring of the implementation process.

Material efficiency approaches cannot be limited to company borders even though this is an important first step. Once the topic is anchored in a critical amount of SMEs, the SC needs to be considered and optimized. The conceptual framework presented by (Gosling, Jia, Gong, and Brown, 2017) may be helpful to introduce material efficiency within SCs. For the future, existing SC networks have to be rethought for a development towards the CE. This change in thinking is associated with several barriers, in particular for SMEs, which have to be overcome (Rizos et al., 2016).

The authors intend to support further research activities with SMEs for the development and implementation of suitable material efficiency tools in enterprises and SCs under consideration of CE aspects. In a first step a proposal for a simplified tool will be developed. Lateron, SC and CE aspects will be attacked, most likely using focal companies.

This investigation was intended to identify tools and methods for the assessment of material efficiency opportunities and to find out which tools are especially suitable for SMEs. The investigation did not include an assessment to which degree the different tools are able to asses the different physical, econonmical and social metrics. At first, it is difficult to assess such a degree in a decent way and secondly it's the metric itself which makes a tool suitable for SMEs and not the characteristic of the metric.

Literature

Allen, D., Bauer, D., Bras, B., Gutowski, T., Murphy, C., Piwonka, T.,...Wolff, E. (2002). Environmentally Benign Manufacturing: Trends in Europe, Japan, and the USA. *Journal of Manufacturing Science and Engineering, 124*(4), 908. https://doi.org/10.1115/1.1505855

Angelakoglou, K., and Gaidajis, G. (2015). A review of methods contributing to the assessment of the environmental sustainability of industrial systems. *Journal of Cleaner Production, 108*, 725-747. https://doi.org/10.1016/j.jclepro.2015.06.094

Bautista-Lazo, S., and Short, T. (2013). Introducing the All Seeing Eye of Business: A model for understanding the nature, impact and potential uses of waste. *Journal of Cleaner Production, 40*, 141-150. https://doi.org/10.1016/j.jclepro.2012.

09.011

Bierer, A., Götze, U., Meynerts, L., and Sygulla, R. (2015). Integrating life cycle costing and life cycle assessment using extended material flow cost accounting. *Journal of Cleaner Production, 108*, 1289-1301. https://doi.org/10.1016/j.jclepro.2014.08.036

Binder, C. R. (2007). From material flow analysis to material flow management Part II: The role of structural agent analysis. *Journal of Cleaner Production, 15*(17), 1605-1617. https://doi.org/10.1016/j.jclepro.2006.08.017

Bleischwitz, R., Welfens, P. J. J., and Zhang, Z. (2012). Prologue to the special issue "Resources, energy and eco-innovation" in Mineral Economics. *Mineral Economics, 24*(2-3), 55-57. https://doi.org/10.1007/s13563-012-0019-8

Boonsothonsatit, K., Kara, S., Ibbotson, S., and Kayis, B. (2015). Development of a Generic decision support system based on multi-Objective Optimisation for Green supply chain network design (GOOG). *Journal of Manufacturing Technology Management, 26*(7), 1069-1084. https://doi.org/10.1108/JMTM-10-2012-0102

Cagno, E., Micheli, G. J., and Trucco, P. (2012). Eco-efficiency for sustainable manufacturing: An extended environmental costing method. *Production Planning and Control, 23*(2-3), 134-144. https://doi.org/10.1080/09537287.2011.591628

Campanale, R., and Femia, A. (2013). An Environmentally Ineffective Way to Increase Resource Productivity: Evidence from the Italian Case on Transferring the Burden Abroad. *Resources, 2*(4), 608-627. https://doi.org/10.3390/resources2040608

Cezar Lucato, W., Vieira Júnior, M., and Carlos da Silva Santos, J. (2013). Measuring the ecoefficiency of a manufacturing process: A conceptual proposal. *Management of Environmental Quality: An International Journal, 24*(6), 755-770. https://doi.org/10.1108/MEQ-10-2012-0063

Chomkhamsri, K., and Pelletier, N. (2011). *Analysis of Existing Environmental Footprint Methodologies for Products and Organizations: Recommendations, Rationale, and Alignment*. Ispra. Retrieved from Joint Research Centre and Institute for Environment and Sustainability website: http://ec.europa.eu/environment/archives/eussd/pdf/Deliverable.pdf

Chung, C.-J., and Wee, H.-M. (2011). Short life-cycle deteriorating product remanufacturing in a green supply chain inventory control system. *International Journal of Production Economics, 129*(1), 195-203. https://doi.org/10.1016/j.ijpe.2010.09.033

Dahlström, K., and Ekins, P. (2007). Combining economic and environmental dimensions: Value chain analysis of UK aluminium flows. *Resources, Conservation and Recycling*, *51*(3), 541-560. https://doi.org/10.1016/j.resconrec.2006.09.010

DeSimone, L. D., and Popoff, F. (1998). *Eco-efficiency: The business link to sustainable development* (2. printing). Cambridge, Mass.: MIT Press.

Devika, K., Jafarian, A., and Nourbakhsh, V. (2014). Designing a sustainable closed-loop supply chain network based on triple bottom line approach: A comparison of metaheuristics hybridization techniques. *European Journal of Operational Research*, *235*(3), 594-615. https://doi.org/10.1016/j.ejor.2013.12.032

DIN-Normenausschuss Grundlagen des Umweltschutzes (NAGUS), Principles of Environmental Protection Standards Committee (Oct. 2006). *Environmental management - Life cycle assessment - Requirements and guidelines*. (DIN EN ISO 14044). Berlin: Beuth.

DIN-Normenausschuss Grundlagen des Umweltschutzes (NAGUS), Principles of Environmental Protection Standards Committee (Nov. 2009). *Environmental management - Life cycle assessment - Principles and framework*. (EN ISO 14040). Berlin: Beuth.

DIN-Normenausschuss Grundlagen des Umweltschutzes (NAGUS), Principles of Environmental Protection Standards Committee (Dec. 2011). *Environmental management - Material flow cost accounting - General framework*. (DIN EN ISO 14051). Berlin: Beuth.

DIN-Normenausschuss Grundlagen des Umweltschutzes (NAGUS), Principles of Environmental Protection Standards Committee (Oct. 2012). *Environmental management - Eco-efficiency assessment of product systems - Principles, requirements and guidelines*. (DIN EN ISO 14045). Berlin: Beuth.

Egilmez, G., Gumus, S., Kucukvar, M., and Tatari, O. (2016). A fuzzy data envelopment analysis framework for dealing with uncertainty impacts of input-output life cycle assessment models on eco-efficiency assessment. *Journal of Cleaner Production*, *129*, 622-636. https://doi.org/10.1016/j.jclepro.2016.03.111

Erlach, K., Köhler, D. C. F., Süchting, M. A., and Kuhlmann, T. (2012). CO2-Wertstrom – Der Weg zur ökonomisch-ökologisch-effizienten Fabrik. *wt Werkstatttechnik online*, *102*(7/8).

European Commission. (2016). Annual report on European SMEs. Retrieved from https://ec.europa.eu/growth/smes/business-friendly-environment/performance-review-2016

Garretson, I. C., Mani, M., Leong, S., Lyons, K. W., and Haapala, K. R. (2016). Terminology to support manufacturing process characterization and assessment for sustainable production. *Journal of Cleaner Production*, *139*, 986-1000. https://doi.org/10.1016/j.jclepro.2016.08.103

Geibler, J. von, Cordaro, F., Kennedy, K., Lettenmeier, M., and Roche, B. (2016).

Integrating resource efficiency in business strategies: A mixed-method approach for environmental life cycle assessment in the single-serve coffee value chain. *Journal of Cleaner Production, 115*, 62-74. https://doi.org/10.1016/j.jclepro.2015.12.052

Giljum, S., Behrens, A., Hinterberger, F., Lutz, C., and Meyer, B. (2008). Modelling scenarios towards a sustainable use of natural resources in Europe. *Environmental Science and Policy, 11*(3), 204-216. https://doi.org/10.1016/j.envsci.2007.07.005

Giurco, D., and Petrie, J. G. (2007). Strategies for reducing the carbon footprint of copper: New technologies, more recycling or demand management? *Minerals Engineering, 20*(9), 842-853. https://doi.org/10.1016/j.mineng.2007.04.014

Gosling, J., Jia, F., Gong, Y., and Brown, S. (2017). The role of supply chain leadership in the learning of sustainable practice: Toward an integrated framework. *Journal of Cleaner Production, 140*, 239-250. https://doi.org/10.1016/j.jclepro.2016.09.101

Graedel, T. E. (1998). *Streamlined Life-cycle Assessment*. New Jersey: Prentice-Hall.

Hassine, H., Barkallah, M., and Bellacicco, A. (2015). Multi Objective Optimization for Sustainable Manufacturing, Application in Turning. *International Journal of Simulation Modelling*, 98-109. https://doi.org/10.2507/IJSIMM14(1)9.292

Hawkins, T., Hendrickson, C., Higgins, C., Matthews, H. S., and Suh, S. (2007). A Mixed-Unit Input-Output Model for Environmental Life-Cycle Assessment and Material Flow Analysis. *Environmental Science and Technology, 41*(3), 1024-1031. https://doi.org/10.1021/es060871u

Herva, M., Franco, A., Ferreiro, S., Alvarez, A., and Roca, E. (2008). An approach for the application of the Ecological Footprint as environmental indicator in the textile sector. *Journal of hazardous materials, 156*(1-3), 478-487. https://doi.org/10.1016/j.jhazmat.2007.12.077

Herva, M., Alvarez, A., and Roca, E. (2012). Combined application of energy and material flow analysis and ecological footprint for the environmental evaluation of a tailoring factory. *Journal of hazardous materials, 237-238*, 231-239. https://doi.org/10.1016/j.jhazmat.2012.08.035

Hoang, V.-N. (2014). Analysis of resource efficiency: a production frontier approach. *Journal of environmental management, 137*, 128-136. https://doi.org/10.1016/j.jenvman.2014.01.046

Ignatenko, O., van Schaik, A., and Reuter, M. A. (2007). Exergy as a tool for evaluation of the resource efficiency of recycling systems. *Minerals Engineering, 20*(9), 862-874. https://doi.org/10.1016/j.mineng.2007.03.005

Jasch, C. (2003). The use of Environmental Management Accounting (EMA) for identifying environmental costs. *Journal of Cleaner Production, 11*(6), 667-676. https://doi.org/10.1016/S0959-6526(02)00107-5

Jayal, A. D., Badurdeen, F., Dillon, O. W., and Jawahir, I. S. (2010). Sustainable manufacturing: Modeling and optimization challenges at the product, process and system levels. *CIRP Journal of Manufacturing Science and Technology, 2*(3), 144-152. https://doi.org/10.1016/j.cirpj.2010.03.006

Kannan, D., Khodaverdi, R., Olfat, L., Jafarian, A., and Diabat, A. (2013). Integrated fuzzy multi criteria decision making method and multi-objective programming approach for supplier selection and order allocation in a green supply chain. *Journal of Cleaner Production, 47*, 355-367. https://doi.org/10.1016/j.jclepro.2013.02.010

Kara, S., Manmek, S., and Kaebernick, H. (2007). An Integrated Methodology to Estimate the External Environmental Costs of Products. *CIRP Annals - Manufacturing Technology, 56*(1), 9-12. https://doi.org/10.1016/j.cirp.2007.05.004

Kluczek, A. (2017). Quick Green Scan: A Methodology for Improving Green Performance in Terms of Manufacturing Processes. *Sustainability, 9*(1), 88. https://doi.org/10.3390/su9010088

Kokubu, K., and Kitada, H. (2015). Material flow cost accounting and existing management perspectives. *Journal of Cleaner Production, 108*, 1279-1288. https://doi.org/10.1016/j.jclepro.2014.08.037

Kubota, F. I., and da Rosa, L. C. (2013). Identification and conception of cleaner production opportunities with the Theory of Inventive Problem Solving. *Journal of Cleaner Production, 47*, 199-210. https://doi.org/10.1016/j.jclepro.2012.07.059

Kuhlmann, T., and Sauer, A. (2016). Ultra-Efficiency Factory - Framework and Holistic Approach to Increase and Evaluate Efficiency and Sustainability. *Advanced Materials Research, 1140*, 481-488. https://doi.org/10.4028/www.scientific.net/AMR.1140.481

Kurdve, M., Shahbazi, S., Wendin, M., Bengtsson, C., and Wiktorsson, M. (2015). Waste flow mapping to improve sustainability of waste management: A case study approach. *Journal of Cleaner Production, 98*, 304-315. https://doi.org/10.1016/j.jclepro.2014.06.076

Latham, G. (2016). Vehicle Remanufacturing: Economic and Environmental Expansion of the Life Cycle. In : *SAE Technical Paper Series*. SAE International400 Commonwealth Drive, Warrendale, PA, United States. https://doi.org/10.4271/2016-01-1291

Latif, H. H., Gopalakrishnan, B., Nimbarte, A., and Currie, K. (2017). Sustainability index development for manufacturing industry. *Sustainable Energy Technologies and Assessments*. Advance online publication. https://doi.org/10.1016/j.

seta.2017.01.010

Lewis, K., and Cassells, S. (2010). Barriers and Drivers for Environmental Practice Uptake in SME's: A New Zealand Perspective. *International Journal of Business Studies, 18*(1), 7-21.

Liedtke, C., Bienge, K., Wiesen, K., Teubler, J., Greiff, K., Lettenmeier, M., and Rohn, H. (2014). Resource Use in the Production and Consumption System–The MIPS Approach. *Resources, 3*(3), 544-574. https://doi.org/10.3390/resources3030544

Loiseau, E., Saikku, L., Antikainen, R., Droste, N., Hansjürgens, B., Pitkänen, K.,...Thomsen, M. (2016). Green economy and related concepts: An overview. *Journal of Cleaner Production, 139*, 361-371. https://doi.org/10.1016/j.jclepro.2016.08.024

Miller, J., Foxon, T., and Sorrell, S. (2016). Exergy Accounting: A Quantitative Comparison of Methods and Implications for Energy-Economy Analysis. *Energies, 9*(11), 947. https://doi.org/10.3390/en9110947

Muñoz, E., Capón-García, E., Laínez, J. M., Espuña, A., and Puigjaner, L. (2013). Considering environmental assessment in an ontological framework for enterprise sustainability. *Journal of Cleaner Production, 47*, 149-164. https://doi.org/10.1016/j.jclepro.2012.11.032

Nakajima, E. S., and Ortega, E. (2016). Carrying capacity using emergy and a new calculation of the ecological footprint. *Ecological Indicators, 60*, 1200-1207. https://doi.org/10.1016/j.ecolind.2015.08.054

Nakamura, S., Kondo, Y., Matsubae, K., Nakajima, K., and Nagasaka, T. (2011). UPIOM: a new tool of MFA and its application to the flow of iron and steel associated with car production. *Environmental science & technology, 45*(3), 1114-1120. https://doi.org/10.1021/es1024299

Nakano, K., and Hirao, M. (2011). Collaborative activity with business partners for improvement of product environmental performance using LCA. *Journal of Cleaner Production, 19*(11), 1189-1197. https://doi.org/10.1016/j.jclepro.2011.03.007

Neugebauer, S., Finkbeiner, Matthias, Volkhausen, Wolfgang, Mecke, S., and Endemann, G. (2013). Environmental evaluation of steel — advanced life cycle assessment considers multiple recycling of steel. *Stahl und Eisen, 133*(7), 49-55.

OECD (2017). *Small, Medium, Strong. Trends in SME Performance and Business Conditions.* Paris: OECD Publishing. Retrieved from http://www.keepeek.com/Digital-Asset-Management/oecd/industry-and-services/small-medium-strong-trends-in-sme-performance-and-business-conditions_9789264275683-en\#.WS3FlevhCUk\#page13

Reuter, M. A., van Schaik, A., and Gediga, J. (2015). Simulation-based design for resource efficiency of metal production and recycling systems: Cases - copper production and recycling, e-waste (LED lamps) and nickel pig iron. *The International Journal of Life Cycle Assessment*, *20*(5), 671-693. https://doi.org/10.1007/s11367-015-0860-4

Rizos, V., Behrens, A., van der Gaast, W., Hofman, E., Ioannou, A., Kafyeke, T.,...Topi, C. (2016). Implementation of Circular Economy Business Models by Small and Medium-Sized Enterprises (SMEs): Barriers and Enablers. *Sustainability*, *8*(11), 1212. https://doi.org/10.3390/su8111212

Rockwell Automation. *Arena Simulation Software*. Wexford. Retrieved from https://www.arenasimulation.com/

Rönnlund, I., Reuter, M., Horn, S., Aho, J., Aho, M., Päällysaho, M.,...Pursula, T. (2016). Eco-efficiency indicator framework implemented in the metallurgical industry: Part 1—a comprehensive view and benchmark. *The International Journal of Life Cycle Assessment*, *21*(10), 1473-1500. https://doi.org/10.1007/s11367-016-1122-9

Saravia-Cortez, A. M., Herva, M., García-Diéguez, C., and Roca, E. (2013). Assessing environmental sustainability of particleboard production process by ecological footprint. *Journal of Cleaner Production*, *52*, 301-308. https://doi.org/10.1016/j.jclepro.2013.02.006

Saurat, M., and Ritthoff, M. (2013). Calculating MIPS 2.0. *Resources*, *2*(4), 581-607. https://doi.org/10.3390/resources2040581

Schaffartzik, A., Wiedenhofer, D., and Fischer-Kowalski, M. (2016). More Productive, Less Sustainable? On the Need to Consider Material Resource Flows. *Intereconomics*, *51*(4), 200-204. https://doi.org/10.1007/s10272-016-0602-2

Schaltegger, S. (2008). *Environmental management accounting for cleaner production. Eco-Efficiency in Industry and Science: Vol. 24*. Dordrecht, Heidelberg: Springer. Retrieved from http://www.springerlink.com/content/xl71r4

Schandl, H., Hatfield-Dodds, S., Wiedmann, T., Geschke, A., Cai, Y., West, J., Owen, A. (2016). Decoupling global environmental pressure and economic growth: Scenarios for energy use, materials use and carbon emissions. *Journal of Cleaner Production*, *132*, 45-56. https://doi.org/10.1016/j.jclepro.2015.06.100

Schliephake, K., Stevens, G., and Clay, S. (2009). Making resources work more efficiently - the importance of supply chain partnerships. *Journal of Cleaner Production*, *17*(14), 1257-1263. https://doi.org/10.1016/j.jclepro.2009.03.020

Schmidt, A., Götze, U., and Sygulla, R. (2015). Extending the scope of Material Flow Cost Accounting - methodical refinements and use case. *Journal of Cleaner Production*, *108*, 1320-1332. https://doi.org/10.1016/j.jclepro.2014.10.039

Sekerka, L. E., and Stimel, D. (2011). How durable is sustainable enterprise? Ecological sustainability meets the reality of tough economic times. *Business Horizons, 54*(2), 115-124. https://doi.org/10.1016/j.bushor.2010.09.006

Simon, F.-G., and Holm, O. (2017). Exergetische Bewertung von Rohstoffen am Beispiel von Kupfer. *Chemie Ingenieur Technik, 89*(1-2), 108-116. https://doi.org/10.1002/cite.201600089

Smith, L., and Ball, P. (2012). Steps towards sustainable manufacturing through modelling material, energy and waste flows. *International Journal of Production Economics, 140*(1), 227-238. https://doi.org/10.1016/j.ijpe.2012.01.036

Steinberger, J. K., Krausmann, F., and Eisenmenger, N. (2010). Global patterns of materials use: A socioeconomic and geophysical analysis. *Ecological Economics, 69*(5), 1148-1158. https://doi.org/10.1016/j.ecolecon.2009.12.009

Steubing, B., Mutel, C., Suter, F., and Hellweg, S. (2016). Streamlining scenario analysis and optimization of key choices in value chains using a modular LCA approach. *The International Journal of Life Cycle Assessment, 21*(4), 510-522. https://doi.org/10.1007/s11367-015-1015-3

Sulong, F., Sulaiman, M., and Norhayati, M. A. (2015). Material Flow Cost Accounting (MFCA) enablers and barriers: The case of a Malaysian small and medium-sized enterprise (SME). *Journal of Cleaner Production, 108*, 1365-1374. https://doi.org/10.1016/j.jclepro.2014.08.038

Sygulla, R., Götze, U., and Bierer, A. (2014). Material Flow Cost Accounting: A Tool for Designing Economically and Ecologically Sustainable Production Processes. In E. Henriques, P. Pecas, and A. Silva (Eds.), *Springer Series in Advanced Manufacturing. Technology and Manufacturing Process Selection* (pp. 105-130). London: Springer London. https://doi.org/10.1007/978-1-4471-5544-7_6

Szargut, J. (2005). *Exergy Method: Technical and Ecological Applications. Developments in Heat Transfer: v.18*. SOUTHAMPTON: WIT Press. Retrieved from http://site.ebrary.com/lib/alltitles/docDetail.action?docID=10621768

Teh, D., and Corbitt, B. (2015). Building sustainability strategy in business. *Journal of Business Strategy, 36*(6), 39-46. https://doi.org/10.1108/JBS-08-2014-0099

Thanki, S. J., and Thakkar, J. J. (2016). Value-value load diagram: A graphical tool for lean-green performance assessment. *Production Planning & Control, 27*(15), 1280-1297. https://doi.org/10.1080/09537287.2016.1220647

Turkay, M., Saracoglu, O., and Arslan, M. C. (2016). Sustainability in Supply Chain Management: Aggregate Planning from Sustainability Perspective. *PloS one, 11*(1), e0147502. https://doi.org/10.1371/journal.pone.0147502

van Berkel, R. (2007). Eco-efficiency in primary metals production: Context, perspectives and methods. *Resources, Conservation and Recycling, 51*(3), 511-540. https://doi.org/10.1016/j.resconrec.2007.03.007

VDI Society Energy and Environment (June 2005). *Systems engineering methodology for design and control of environmental relevant processes in the operational area - General basics*. (VDI 4090 part 1). Berlin: Beuth.

VDI Society Energy and Environment (2012, January 1). *Cumulative energy demand (KEA) - Terms, definitions, methods of calculation*. (Richtlinie, 4600). Berlin: Beuth.

VDI Society Energy and Environment (October 2014). *Cleaner production (PIUS) - Basic principles and area of application*. (Guideline, 4075). Berlin: Beuth.

VDI Society Energy and Environment (2016a, February 1). *Resource efficiency - Methodical principles and strategies*. (Richtlinie, 4800 part 1). Berlin: Beuth.

VDI Society Energy and Environment (2016b, March 1). *Resource efficiency - Evaluation of the use of raw materials*. (Richtlinie, 4800 part 2 Entwurf). Berlin: Beuth.

Wall, G. (1977). *Exergy - A Useful Concept within Resource Accounting* (Report no. 77-42,). Goteborg Sweden.

Wang, X., Dadouma, A., Chen, Y., Sui, P., Gao, W., and Jia, L. (2015). Sustainability evaluation of the large-scale pig farming system in North China: An emergy analysis based on life cycle assessment. *Journal of Cleaner Production*, *102*, 144-164. https://doi.org/10.1016/j.jclepro.2015.04.071

Wiedmann, T. O., Schandl, H., and Moran, D. (2015). The footprint of using metals: New metrics of consumption and productivity. *Environmental Economics and Policy Studies*, *17*(3), 369-388. https://doi.org/10.1007/s10018-014-0085-y

Xue, H., Kumar, V., and Sutherland, J. W. (2007). Material flows and environmental impacts of manufacturing systems via aggregated input-output models. *Journal of Cleaner Production*, *15*(13-14), 1349-1358. https://doi.org/10.1016/j.jclepro.2006.07.007

Yang, S., Yang, S., and Qian, Y. (2015). The inclusion of economic and environmental factors in the ecological cumulative exergy consumption analysis of industrial processes. *Journal of Cleaner Production*, *108*, 1019-1027. https://doi.org/10.1016/j.jclepro.2015.07.039

Zhao, R. (2012). Simulation-Based Environmental Cost Analysis for Work-in-Process. *International Journal of Simulation Modelling*, *11*(4), 211-224. https://doi.org/10.2507/IJSIMM11(4)4.218

Zhong, S., Geng, Y., Liu, W., Gao, C., and Chen, W. (2016). A bibliometric review on natural resource accounting during 1995-2014. *Journal of Cleaner Production*, *139*, 122-132. https://doi.org/10.1016/j.jclepro.2016.08.039

Zhou, Z., Zhao, W., Chen, X., and Zeng, H. (2017). MFCA extension from a circular economy perspective: Model modifications and case study. *Journal of Cleaner Production*, *149*, 110-125. https://doi.org/10.1016/j.jclepro.2017.02.049

A Study on Challenges in the Implementation of Human-Robot Collaboration

Christopher Gaede[a], Fabian Ranz[a], Vera Hummel[a], Wolfgang Echelmeyer[a]

[a] *ESB Business School, Reutlingen University, D-72762 Reutlingen*

Abstract

While academia and industry see large potential for Human-Robot-Collaboration (HRC), only a small number of realized HRC applications is currently found in industry. To gather more data about current hindrances to wider implementation of collaborative robots, a study among 15 robot manufacturers and 14 system integrators of collaborative robot technology has been conducted through a predesigned questionnaire procedure. Additionally, five industrial users of human-robot collaboration have been interviewed on the main challenges they experienced during the initial implementation process. The quantitative data has been analyzed using the Wilcoxon-Signed-Rank-Test. According to the study participants, the main challenges within the implementation currently are the identification of HRC-suitable processes, the application of relevant safety norms (such as ISO 10218, ISO/TS 15066) and the application-individual risk assessment.

1 Introduction

Current economic challenges such as the demographic change, high labor costs, rising product individualization and shorter innovation cycles force the manufacturing industry to make use of innovative production principles and technologies in order to maintain value-creation both cost-effective and flexible. The socio-technical concept of human-robot-collaboration as well as late technological developments in the area of humanoid robots promise the opportunity to improve the ergonomics of work processes and increase the productivity by flexibly automating suitable processes and make use of robots' and humans' strengths at the same time. Although the scenario has already been existing for decades (Colgate et al., 1996), robots specifically designed for HRC became available only within recent years. While the market potential for this new generation of robots can - due to more intuitive programming and comparably low procurement cost - be considered high, only a limited number of actual examples of successful human-robot collaboration is currently to be found in the industry. The following study was conducted in order to gather a comprehensive understanding about the current challenges and barriers to the implementation of HRC. From the study results, auspicious research areas and questions can be derived.

2 Theoretical background and definitions

2.1 Hybrid working systems

Hybrid working systems represent a combination of manual and automated processes in which humans and machines work together (Lotter and Wiendahl, 2012). While automated systems are beneficial for high volume/low variance production, manual systems are rather favorable in low volume/high variance environments due to the human capability of adapting to frequently changing work requirements. Hybrid working systems offer a bridge between these two options and combine human capabilities in problem solving, creativity, adaption and anticipation, robots have their advantages through a high operating speed, repeated accuracy and endurance (Michalos et al., 2014). By applying the different strenghts of humans and robots in one working system, productivity can be improved compared to purely manual systems while a higher degree of flexibility than within hard automation can be maintained.

2.2 Human-robot collaboration

Human-robot collaboration is defined as a "state in which a purposely designed robot system and an operator work within a collaborative workspace" (ISO 10218-1:2011, 3.4). The collaborative workspace is further defined as the "space within the operating space where the robot system (including the workpiece) and a human can perform tasks concurrently during production operation" (ISO 10218-1:2011, 3.5). In practice, HRC describes a situation where humans and robots work jointly on a certain task within a narrow workspace and under absence of safety fences. In order to ensure safety in such, ISO 10218-1 proposes four possible operational concepts for HRC.

Within *safety-rated monitored stop*, the robot decreases its velocity or stops its movements when a person enters the collaborative workspace and continues its movement only after the person has left the workspace again. In *hand guiding*, the human directs the robot through manual force input while the robot only passively provides its lifting power to the worker, for example in moving heavy objects. *Speed and separation monitoring* describes a scenario in which the robot constantly determines the position of the worker in the collaborative workspace and ensures that a certain safety distance to the worker is maintained in order to prevent a collision. Only in *power and force limiting* collaboration mode, human and robot move independently and side by side. The risk of an actual collision is accepted here, under the precondition that potential forces impacting the human operator during this collision do not exceed threshold values as for example provided by ISO/TS 15066. If application safety is achieved, human-robot collaboration can provide certain benefits to human operators. By allocating ergonomically adverse tasks to the robot, HRC can reduce work-related burden (Busch et al.,

2013). During workload spikes, implementation of a robot into a production system can add certain extra capacity (Michalos et al., 2014). While standard industrial robots have been implemented in HRC applications such as (Thiemermann & Schraft, 2003), power and force limited robots especially fitted for this last collaboration mode have been introduced to the market since 2009. They monitor motor currents or are equipped with force torque sensors to recognize collisions with their environment. Mostly, articulated robots are used for the purpose of HRC.

3 Research methodology

3.1 Research approach

In order to identify the current challenges in implementing human-robot collaboration data was collected through a survey among companies with a pre-designed questionnaire. Furthermore, in-depth interviews were conducted to verify the results of the survey and gather qualitative data about the challenges in practice.

The predesigned questionnaire consisted of two parts: The first part investigated the opinions on future potential of HRC as well as possible fields of application. The second part focused on current challenges in implementation. A variety of possible challenges in the implementation process was listed and interviewees were asked to evaluate the practical relevance and importance of the individual challenges and obstacles. Throughout the survey, a four-staged Likert-Skala was used. The obtained data was analyzed by the Wilcoxon-Signed-Rank-Test (Fahrmeier et al., 2007) which allows to establish a ranking of all identified challenges. Three external experts have been consulted prior to the survey to ensure the list of challenges suggested to the interviewees is complete. Also, each interviewee could add additional challenges to the questionnaire he experienced in his respective projects.

The in-depth interviews were undertaken via telephone. The guideline for the interview was constructed with respect to the questionnaire in order to obtain coherent data. The qualitative data was compared with the results of the survey to be able to identify accordances or differences in opinions.

3.2 Sample structure

For the identification of the current challenges in the implementation process of HRC applications, the opinions of the three main groups participating in the implementation were examined. These are:

Type of participant	Number of evaluated participants
Manufacturer of robot technology[1]	15
System integrator of collaborative robots[1]	14
User of collaborative robots	5

The five *users* of collaborative robots that have been interviewed as part of the second study phase belong to the following industries:

#	Type of HRC user
1	Combustion engine manufacturer
2	Car manufacturer
3	Power tools manufacturer
4	Household appliance manufacturer
5	Cleaning equipment manufacturer

The robot manufacturers and system integrators of collaborative robots were surveyed at the industrial fairs "Hannover Messe" in Hannover and "Automatica" in Munich. The opinions of the users of collaborative robots were gathered through in-depth interviews.

4 Results of the survey

4.1 Potential of collaborative robots

The future potential of human robot collaboration was regarded large or very large by 93% of robot manufactures and likewise 93% of the system integrators. The five HRC users also expected a large to very large potential, with one user having >25 HRC applications operating in European factories. Furthermore, no participants rated the future potential as very small. The data shows that both producers and system integrators expect a large future potential for collaborative robots.

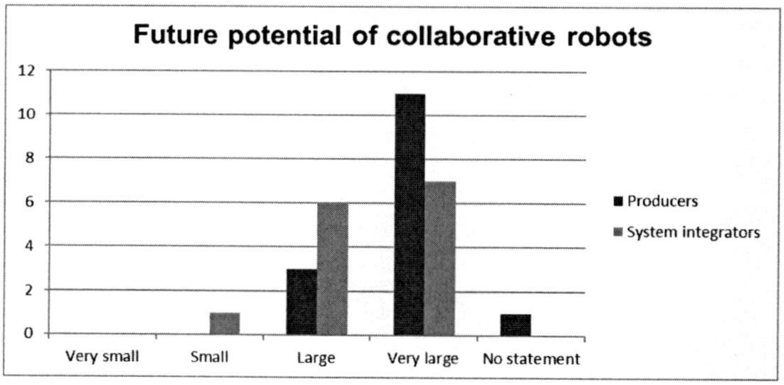

Figure 1: Subjective rating of HRC potential

[1]*Manufacturers* are those that develop, produce and market robots, but are only rarely involved in system implementation. *System integrators* are defined as companies that specialize in implementing automation systems and also offer the implementation of collaborative robots, but do not manufacture robots or components.

The surveyed robot manufacturers of collaborative robot technology rated the interests of different company sizes in implementing collaborative robots rising with growing size. The interest of small companies (< 50 employees) was expected lower by the participants than the interest of medium (< 250 employees), large (< 1000 employees) and very large (> 1000 employees) companies. While 53% of the producers expected a large or very large interest of small companies for collaborative robots, the figure increased to 80% with regard to very large companies.

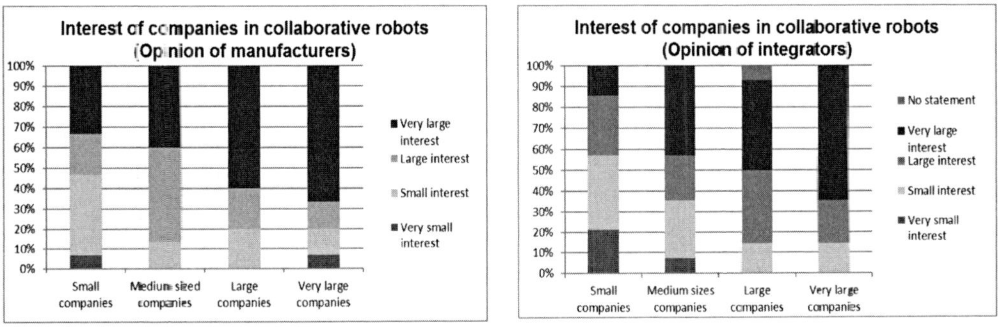

Figure 2: Interest of companies in HRC per company size

The system integrators surveyed have a similar opinion compared to the robot manufacturers and rate the interest of smaller companies in HRC lower than the interest of larger companies (see Fig. 2). The percentage of system integrators rating the interest of small companies large or very large was 43% while it was 86% for very large companies. While the authors expected more intense interest in HRC by small companies due to the comparably low baseline cost of power and force limited robots, the stronger interest of larger companies can probably be explained by their broader experience with and affinity to robots in general.

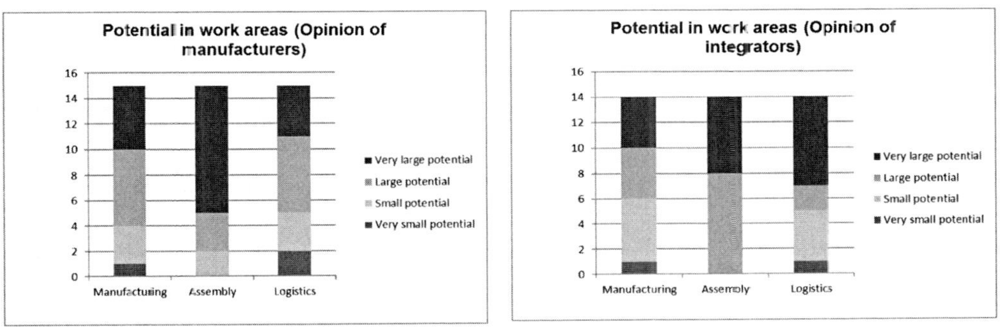

Figure 3: Rated potential of HRC per process area

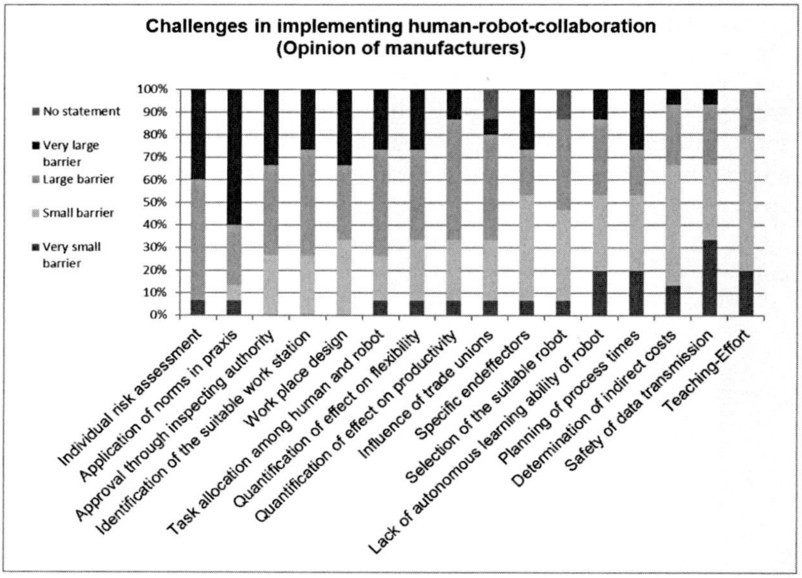

Figure 4: Challenges in HRC implementation, as perceived by robot manufacturers

Robot manufacturers see the highest potential for implementing HRC in assembly processes, where 87% voted for a large or very large potential (see Fig. 3). The potential for manufacturing and logistics was rated lower. However, 73% still vote for a large to very large potential in manufacturing and 67% for a large to very large potential in logistics. This result was predicted by the authors due to the fact that assembly systems are still predominantly manual work systems (Spena et al., 2016) with mostly repetitive tasks that are favorable for robotic execution, while manufacturing systems are already automated to large extent using traditional automation methods.

Similarly, the system integrators see the highest potential in the assembly where all participants voted for a large or very large potential. Furthermore, 57% of the participants voted for a large to very large potential in manufacturing and 64% for a large to very large potential in logistics.

4.2 Challenges in the implementation of collaborative robots

Figures 4 and 5 outline the relevance of 16 individual challenges that can occur when implementing HRC applications as considered by robot manufacturers (figure 4) and system integrators (figure 5).

The suggested challenges are categorized into safety-related challenges, organizational-/process-related challenges and technical challenges, which mostly deal with technical deficiencies of the available robotic products. The top five challenges iden-

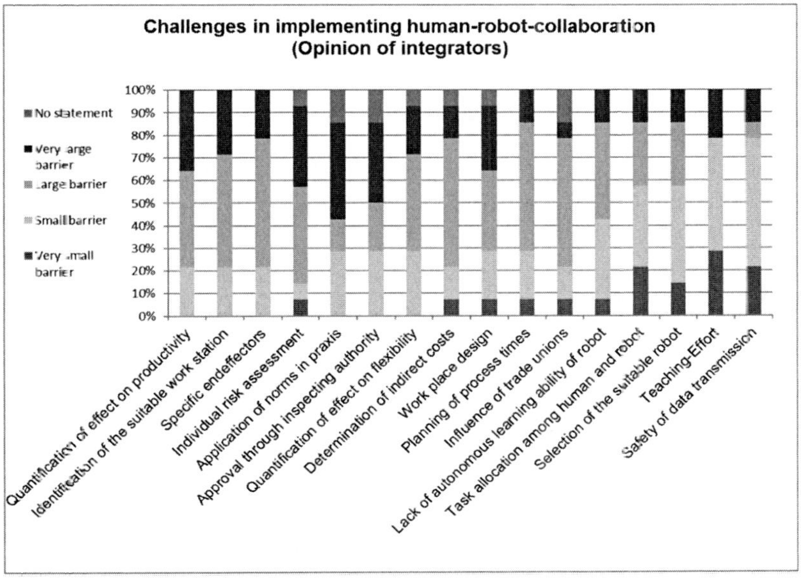

Figure 5: Challenges in HRC implementation, as perceived by system integrators

tified by both robot manufacturers and system integrators will be outlined and discussed taking into account the results from the in-depth user interviews.

The data shows that the surveyed robot manufacturers as well as the surveyed system integrators agree on three challenges that they both ranked within the top five. These challenges are: the *identification of HRC-suitable workstations*, the *application of the relevant safety norms* such as ISO 10218-2 or ISO/TS 15066 as well as the *conduction of the required application-individual risk assessment*.

Firstly, the identification of the suitable workstation was voted as a large or very large challenge by 73% of the manufacturers and 79% of the system integrators. Three out of five HRC users made their lack of experience accountable for the difficulties in identifying workstations that offer potential for meaningful HRC, stating that no guidelines or methods that support this task are available so far. One user for instance, a German car OEM, went through three ineffective HRC applications in a "try-and-error" manner before finding one that was feasible from both technical as well as organizational standpoint. One user, a power tools manufacturer, blamed poor automation capacity of his products for the difficulties in identifying feasible HRC applications and will focus stronger on automation capacity in future product developments.

The application of the relevant safety norms is regarded as a large or very large challenge by 87% of the robot manufacturers, but only 57% of the system integrators, probably based on the fact that primarily robot manufacturers are responsible for delivering safe hardware that is generally compliant with the wide range of existing norms and standards. Furthermore, for 73% of the robot manufacturers and 79% of the system integrators the individual risk assessment constituted a large or very large problem.

During the in-depth interviews, also all of the actual users of collaborative robots confirmed that assuring compliance with the relatively new safety norms for HRC and conduction of a risk assessment on HRC applications happen to be a barrier in HRC implementation. They again blame lack of internal experience and of available practical guidelines that could support the design and implementation phase. Also, the discrepancy of the safety levels between the actual robot arm, which might be inherently safe, and the end-effector, which might not be fitted with same safety features, complicate application design and safety assurance.

Beyond that, 73% of robot manufacturers perceive the required re-certification of a HRC-application by an inspection authority every time the application is changed as a large or very large barrier towards further dissemination of HRC as it compromises on the short-cycled flexibility many collaborative robots offer, and causes additional bureaucratic effort. However, four out of five HRC-users stated in the interview that this perception is only true to fundamental changes in the application whereas minor adjustments can get re-certified with low effort.

79% of systems integrators consider difficulties in quantification of productivity effects through HRC a large or very large barrier to HRC implementation, whereas only 67% of robot manufacturers agree. This aspect is perceived very differently by the interviewed users of HRC applications: Two out of five users found it easy to predict and measure the productivity of the hybrid work place. A third user did not focus on improving the productivity through collaborative robots, but stated that the planning of process times of the hybrid work station required a high effort, adding: "currently we are only trying to get applications running in order to gain experience. Actual productivity is secondary for now." Another user was able to measure improvements in quality and a reduction of sick days, but similarly found the planning of the process times a difficult task due to a missing planning tool for hybrid working systems, while one user said that the incapacity to approximate productivity effects upfront even stopped an company-initial HRC initiative in 2014.

Interestingly, also 79% of the system integrators ranked the high specificity/low flexibility of grippers and end-effectors a very large or large barrier to implementation of collaborative robots, whereas only 45% of equipment manufacturers have a corresponding perception - a clear indication that robot manufacturers are not fully aware of market requirements in terms of effectors. Also, all five users consider inflexibility of tools a current barrier that cuts back on the inherent flexibility of modern robots. One user suggests assembly process sequences could be re-arranged to better utilize specific tools by "centralizing" all those process steps that have same end-effector requirements.

Work place design for HRC, perceived a very large or large challenge by 68% of robot manufacturers and 64% of system integrators, was rated differently by the interviewed users. Three out of five found this aspect challenging due to the difficulty of realizing the technical design, for example the material staging, especially since there is no industry standard at hand like for manual work places. However, the remaining two users did not perceive the work place design as a major challenge in their application.

4.3 Differences in opinions between robot manufacturers or system integrators and users

Within the study, the opinions between robot manufacturers or systems integrators and users diverged. Three essential findings will be outlined.

Firstly, the effort for teaching the collaborative robot was seen as a small or non-relevant challenge as 80% of the producers and 79% of the system integrators perceived it as a small or very small barrier. However, the users stated in the interview that the effort for teaching fluctuates depending on the type and brand of selected robot for the HRC application. Within some user companies, teaching and minor maintenance is already undertaken by operators themselves, while other user companies need specialist staff for programming, which hampers both flexibility and profitability of the application.

Secondly, the selection of the suitable robot is regarded a small challenge by 80% of the surveyed robot manufacturers but only 57% of the system integrators. Similarly, two users described the selection as a difficult process due to a non-sufficient cost-benefit ratio or missing information about safety features and risk classifications.

Thirdly, the allocation of tasks between human and robot in a hybrid working system was placed on rank 6 by producers and rank 13 by system integrators which represents a difference of 7 ranks. All users perceived the allocation as a large challenge due to different reasons. These reasons are primarily a lack of experience as well as a limitation of possible task arrangements due to either the speed limit of collaborative robots or safety requirements.

Fourthly, the answers of the interviewed users on the determination of process cycle times of the hybrid working system show different states of knowledge. While two users perceived the task as easy and non-challenging, two users described the determination of cycle times as challenging due to a missing planning tool which takes into account human and robot movements in one simulation. Furthermore, one of these users missed sufficient information about the motion paths of the robot which complicated the planning of process times.

5 Conclusion

The vast majority of the study participants expects a large future potential for human robot collaboration especially in larger firms and mainly within assembly process areas.

However, current challenges and barriers still hamper the implementation of human-robot collaboration as a manufacturing concept although capable technology is available for years. Robot manufacturers and system integrators both see the main challenges in the identification of HRC-suitable workstations, the application of the relevant safety norms and the difficulties in implementing the application-individual risk assessment. In the interviews, the users confirmed this perspective and emphasized the difficulty of ensuring safety under consideration of specific end-effectors and work pieces. Other challenges like the high specificity of grippers or the re-certification of the collaborative application after an application change only appear to be a problem when a very flexible use of the robot is envisaged.

In general, the study proves that safety-related aspects as well as process design-related aspects primarily impede HRC dissemination in the industry, whereas technical aspects such as equipment selection, teaching effort or lack of machine intelligence play a subordinate role.

In order to simplify the implementation process of collaborative robots, several future measures could be pursued. Universities and research institutes could play a leading role in facilitating future distribution of HRC into industrial environments by developing methods and procedures which address the challenges outlined in this study, such as risk assessment methods specifically for HRC, structured workplace identification approaches, generic HRC workstation design templates or structured patterns to quantify productivity and profitability of HRC applications upfront. On the other hand, industrial companies are required to go ahead and implement pilot applications to gain practical knowledge - a missing link between available technology for HRC and existing industrial processes as this study revealed. In a combination of structured and validated implementation approaches and a higher degree of experience on the user side, HRC could reach a far more extensive spread within the years to come.

Acknowledgements

The authors thank all study participants for contributing to this survey with their experiences and thereby creating valuable knowledge that led to this publication.

References

Busch F, Hartung J, Thomas C, Wischniewski S, Deuse J, Kuhlenkötter B (2013). Individualisierte Arbeitsassistenz in der Produktion - Gestaltung gesunder, sicherer und wettbewerbsfähiger Arbeit in der industriellen Produktion. Industrie Management Vol. 29 (2013), pp. 2-10.

Colgate JE, Wannasuphoprasit W, Peshkin M (1996). Cobots Robots for Collaboration with Human Operator. Proceedings of the ASME Dynamic Systems and Control Division, DSC-Vol. 58, 433-440.

Fahrmeier L, Künstler R, Pigeot I, Tutz G (2007). Statistik - Der Weg zur Datenanalyse. 6th edition. ISBN 978-4-540-69713-8, Springer, Heidelberg, 2007.

Lotter B, Wiendahl HP (2012). Montage in der industriellen Produktion - ein Handbuch für die Praxis. 2nd edition. ISBN 978-3-642-29060-2, Springer, Heidelberg,

2012.

Michalos G, Makris S, Spiliotopoulos J, Misios I, Tsarouchi P, Chryssolouris G (2014). ROBO-PARTNER: Seamless Human-Robot Cooperation for Intelligent, Flexible and Safe Operations in the Assembly Factories of the Future. Procedia CIRP 23 (2014) pp. 71-76.

Spena, PR, Holzner, P, Rauch, E, Vidoni, R, Matt DT (2016). Requirements for the Design of flexible and changeable Manufacturing and Assembly Systems: a SME-survey. Procedia CIRP 41 (2016) pp. 207-212.

Thiemermann S, Schraft RD (2003). team@work–Mensch-Roboter-Kooperation in der Montage. Automatisierungstechnische Praxis atp Praxis der Mess-, Steuerungs-, Regelungs- und Informationstechnik, vol. 45(11). pp. 31-35.

Digital Twins in the Smart Factory

Marius Greger[a], Nur Budi Mulyono[b], Julian Hermann[a], Daniel Palm[a]

[a] *ESB Business School, Reutlingen University, D-72762 Reutlingen*

[b] *School of Business and Management, Institut Teknologi Bandung, 40116 Bandung, Indonesia*

Abstract

What does the factory of tomorrow have to offer for companies? This question and its aspects are the focus of many actual articles and publications. According to Gartner digital twins, one of 2017 strategic technology trends will play a big role for the future of manufacturing. At the moment digital twins are gaining more importance for the industrial application. If companies want to be competitive in the future they have to implement the digital twin in the factories of today. Therefore this paper provides a basic overview of the concept of the smart factory and its requirements. In addition, digital twins are identified as a necessary concept for the evolution of the factory of today.

1 Introduction

History shows that an industrial revolution always leads to economic growth (Scalabre, 2016). Actually three industrial revolutions have taken place so far. The first one took place with the development of the steam engine in the middle of the 19th century. In the beginning of the 20th century Henry Ford's idea of the mass production model (piecework and the assembly line) led to the second industrial revolution. The automation of manufacturing processes in the 1970s led to the ongoing third industrial revolution (Vogel-Heuser, Bauernhansl and ten Hompel, 2017, pp. 1–4).

Those three revolutions had one thing in common: they enabled an economic growth by an increase of the productivity. Based on these facts, the following hypothesis can be built: "In order to generate economic growth it is necessary to first increase productivity". As figure 1 shows, quite the opposite is happening at the moment, as the productivity has been slowly decreasing.

Many experts claim, that there is a big change in the industry happening at the moment, especially the manufacturing industry is said to face substantial changes within the next years (Pfeiffer et al., 2016). These changes mark the beginning of the fourth industrial revolution, named "Industry 4.0" by a German initiative. Compared to the

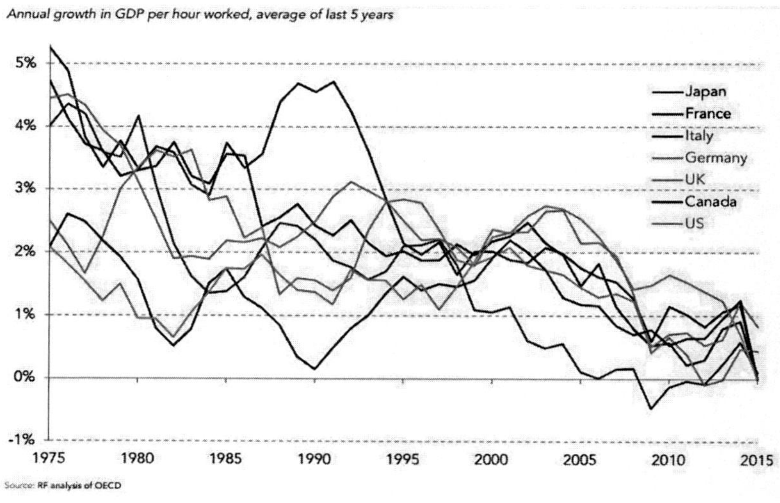

Figure 1: Anual growth in GDP per hour worked, average of last 5 years (Sherman, 2017). The diagram shows the slow decrease of the productivity growth between 1975 and 2015. Therefore the productivity of Japan, France, Italy, Germany, UK, Canada and the US is compared.

attempts of the last 30 years, the fourth industrial revolution has to generate an increase in productivity again. The attempts carried out so far have not led to any decisive increases in productivity yet (Sirkin, Zinser and Rose, 2015; O. Scalabre, 2016). For example, many companies attempted to cut costs by building their production facilities offshore while benefitting from low labour costs there. However, this strategy does not led to higher productivity – it was just changing the place of value creation. The big problem with this strategy is that it only leads to saving of costs in a short time, because labour costs are increasing steadily all over the world. Moreover, this could create unemployment and might cause the loss of customers in the company's home country, since such strategies are seen unethical and not supporting the local activities. Another strategy was to build large factories that are specialized for specific products. The idea was to manufacture a huge amount of one product and stockpile it to be sold on demand. This attempt increased the productivity for a short while, but it introduced many rigidities in supply chains. Nowadays, both the market and the customer have changed. The demand of individual/customized products is increasing, and therefore this approach is not applicable any more (Scalabre, 2016).

A big problem is that most of the factories remained the same as 30 years ago, while their surroundings have completely transformed. Therefore, it is important for the manufacturing industry to adopt new technologies and approaches to increase the productivity of the factories of today (Scalabre, 2016). Currently the invention of new technologies and concepts are happening under the name Industry 4.0.

Industry 4.0 deals with the digitalization of the entire value-chain, from the extraction of raw materials through the entire lifecycle of products. Furthermore, Industry 4.0 deals with concepts like the connection and autonomous communication of machinery, the digital twin, the smart factory, predictive maintenance, big data, etc. (Plattform Industrie 4.0, 2017). This paper deals with the concept of the digital twin in the smart factory.

2 Smart Factory

The future of factories is a part of the high-tech-strategy "Industry 4.0" initiated by the German Ministry of Education and Research (Bundesministerium für Bildung und Forschung, 2017). In this frame, the *smart factory* and *digital manufacturing* are seen as the future of the factories of today (Bundesministerium für Bildung und Forschung, 2017).

The smart factory is understood as a vision of the manufacturing environment in which production plants and logistics systems are working self-organized. Another feature of the smart factory is the ability to produce intelligently and autonomously. The technological basis for the smart factory was set with the implementation of information and communication technology (ICT) in the third industrial revolution. ICT is primarily used in the manufacturing environment to optimize costs and efficiency. Other fields of use are in logistics, quality management, maintenance, energy management and for employee resource planning. ICT also plays an important role for the digital- and virtual factory (Zillmann and Wilk, 2016, p. 9).

A common mistake is that the smart factory is used as a synonym for the digital- or virtual factory. While the digital factory focuses on the software for the digitalization of products and processes, the virtual factory focuses on the connectivity within the supply chain. Compared to those two, the smart factory focuses on the optimization and management of the whole production process (see figure 2) (Zillmann and Wilk, 2016, pp. 9–10).

The crucial parts of the smart factory are those ICT-Systems, which are used for the dynamic management of complex production systems, and make the real time intelligent horizontal and vertical linkage of humans, machines and objects possible. Therefore, it is important that all items of a factory – starting from single tools to products and machineries – are able to collect, understand and share data out of their environment. This data is shared with all the management systems like the supply chain management (SCM), manufacturing execution system (MES), enterprise resource planning and product lifecycle management (Zillmann and Wilk, 2016, p. 14).

A challenge of the smart factory are the IT-systems, which have grown over the years. Caused by their complexity and closed architectures, the exchange of information between those systems is combined with a high effort. Therefore, a link between those IT-systems as well as a flexible approach for the reconfiguration of production systems is often not realized and a lot of potential cannot be used. This conditions will change due to the implementation of Industry 4.0 (Kagermann, Wahlster and Helbing, 2013).

Stages of ICT- Based Manufacturing

	Digital Factory	Smart Factory	Virtual Factory
Purpose	Digital pre-design of products, Factory- and construction planning	Automation, Optimization and Management of the production	Management of supply chains, value add by integrating products and services
Realization	Software for the digitalization of products and production processes	ICT, Factory infrastructure, Embedded cyber-physical items	ICT for connectivity, Management of production Capacity, New business models
Expected	Design Productivity	Manufacturing	Supply Chain

Figure 2: Stages ICT-based production (Zillmann and Wilk, 2016, p. 10). The table points out the differences between the terms digital factory, smart factory and virtual factory.

The advantages of the smart factory for companies can be summarized as (Zillmann and Wilk, 2016, p. 40):

Total Transparency: all processes are captured from the assembly lines to the management processes.

Predictive Management: the analysis of big data in real time allows companies to react fast and flexible to changes.

Autonomy: thanks to an increased degree of automation, production plants and logistics systems work mostly together and autonomously.

Lot size one: individualized products can also be produced economically in the serial production.

Flexibility: sequential supply chains will be replaced by flexible and international value chains.

Globalization: manufacturers and suppliers will be able to increase their productivity and their degree of innovation.

2.1 Spread of the Smart Factory

Based on the actual study of Lünedonk "Der Weg zur Industrie 4.0 – Wie Verantwortliche aus den Bereichen SCM und Produktion den Status quo beurteilen" the term

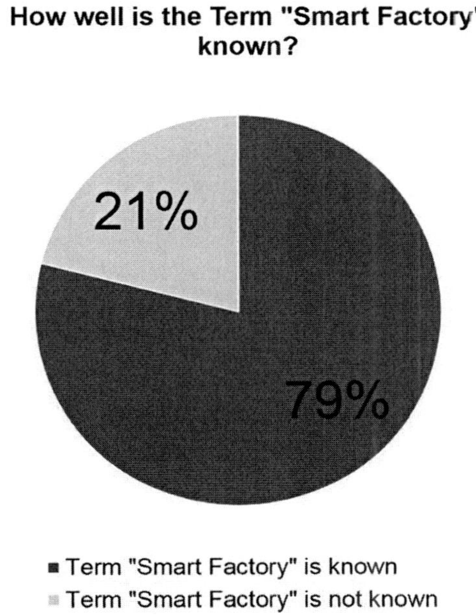

Figure 3: Popularity of the Smart Factory (Zillmann and Wilk, 2016, p. 10). This figure displays how much percentage of the managers out of the area of SCM and manufacturing are familiar with the concept of the smart factory

smart factory is already well known, as displayed in Figure 3.

Therefore, the increase of digitalization in the factories leads to the implementation of the smart factory. The question is: how fast will this implementation happen? At the moment, there is no answer to this question, because the opinions are contrary. While there are many speculations about the speed of development, there are several institutes which recently researched the *digital readiness* of companies. The Foundation for Mechanical Engineering, Plant Engineering and Information Technology focused its research on the area of manufacturing industry. Their study concludes that the readiness of the manufacturing industry is very low compared to other industrial sectors. Sectors like the insurance industry, banking or the travel industry, are completely data driven and the digitalization of their processes started earlier in the past and is nowadays an essential part of their structure. The sectors, which own big facilities like industrial manufacturing or energy producers are in the back spots of Industry 4.0 at the moment and have to catch up (Zillmann and Wilk, 2016, pp. 15–16).

The previously mentioned fact is supported by the "Industrie 4.0 Readiness IMPULS Studie", which shows that 95,3 percent of the participating companies in 2015 are only in stages 0–2 out of 5 in becoming a smart factory. This means that they either won't or barely meet the requirements to be called a smart factory (Lichtblau et al.,

2013). Some of the main reasons for this situation are (Zillmann and Wilk, 2016, p. 26):

- The risk of production delays (or downtime) caused by changes in the working production systems.
- High costs in manufacturing. New investments are only started if old investments have amortized. Therefore, the manufacturing industry prefers to make many small investments rather than a big one.
- The lack of ports or flexibility in the present software solutions.
- The different worldviews of the stakeholders.

To conclude, the factory of today has to transform into the smart factory. On the way to the smart factory many obstacles have to be overcome to make the smart factory adoptable in the future.

2.2 Requirements of the Smart Factory

Before the relevance of the digital twin in the smart factory can be discussed, the requirements of the smart factory need to be evaluated. Based on these requirements the link of the smart factory and the digital twin will be discussed in section 4. Dr. Alexander Schließmann, Director and Chief Technology Officer of Forcam already defined 10 requirements for the smart factory in 2015, which are as following (Schließman, 2015):

Connectivity: Most factories consist of heterogeneous machine parks. The task of manufacturing IT is to upgrade those machine parks so they are able to collect and share their data (heterogeneous, because they are often from different brands and made in different years). Therefore, it is necessary to create one platform, where they can communicate.

Availability: The shopfloor-technology has to be web based so that all users and all devices can communicate worldwide. The data and the applications could work on one central server in the cloud. Therefore, users would only need a device that is equipped with a browser.

Security: In the manufacturing industry, the security of the production data is an essential part for the acceptance of models. Therefore, companies will tend to build their own cloud, called "private cloud". In the future, the need of "hybrid-clouds" will raise because companies can cut down their IT-architecture-costs by using them.

Speed: In order to analyze and visualize large volumes of data in real time a high performance is needed. Therefore, companies already use a combination of memory-based in-memory technology and complex event processing (CEP). This provides access times that are 100 times higher than traditional approaches.

The CEP technology comes from the high-frequency trade of the financial world and enables complex links, subsequent adjustments and repeated interpretations.

Cyber-Reality: The aim of Industry 4.0 is to generate a cyber-physical system (CPS), which is a digital twin of the real and physical production area in real-time. With a twin like that, companies can analyze and optimize their processes. Therefore, an infrastructure that is reflecting the whole production is needed to track all processes and products for the fast identification and solving of problems.

Flexibility and Variability: Production cycles vary due to shifts or interruptions which take place during the changeover or set-up. The costs increase exponentially the longer a re-dispensing or a disturbance correction does not occur. Modern manufacturing systems react to this by flexibly configuring and using them variably. This applies in particular to production across national boundaries when the IT works with different computer structures for performance or security reasons. Therefore modern shop floor solutions can work together with central servers, distributed servers and mixed forms.

Globality: A globally oriented production software should be able to display and visualize master and movement data for all end user devices in all languages. This can be mastered by storing text elements in the database in advance and make them available for translation. In addition, handling of all foreign characters should be supported.

Time: A time difference of one second between a quantity report and an order can lead to rejected bookings because they are supposed to be in the future. In order to prevent such errors, modern MES solutions should be based on two levels: Firstly, time events can be recorded as an UTC time code so that the data of different time zones can be compared. On the second level, the determined time stamps can be compared worldwide. The pre-set times are in all computers and software in parts, machines and products. At the moment, many companies try to evaluate new use cases for the blockchain. This technology could also be applied to solve the aspect of time.

Visualization: The visualization of the software platform of the smart factory is very important because it could lead to a high transparency and acceptance. Moreover, fast and meaningful visualizations allow fast decision-making for each role and function during operations.

Acceptance: All workers should be integrated in the transformation of the production from early stages on and be informed about the progress constantly. It is recommended to find a test environment for the implementation and to build some rules for a successful rollout for a later stage.

3 The Digital Twin

The digital twin was first announced by NASA to the public in 2010. According to NASA's implementation of a digital twin, they describe their model as "an integrated multiphysics, multiscale simulation of a vehicle or system that uses the best available physical models, sensor updates, fleet history [...] to mirror the life of its corresponding flying twin" (Shafto et al., 2010; Shafto et al., 2012).

At the end of 2016, Gartner predicted digital twins to be one of the top 10 strategic technological trends for 2017. Gartner is one of the word leading research and advisory companies. The digital twin is defined by Gartner as: "a dynamic software model of a physical thing or system. Using physics data on how the components of a thing operate and respond to the environment as well as data provided by sensors in the physical world, a digital twin can be used to analyze and simulate real world conditions, responds to changes, improve operations and add value. Digital twins function as proxies for the combination of skilled individuals (e.g., technicians) and traditional monitoring devices and controls (e.g., pressure gauges). Their proliferation will require a cultural change, as those who understand the maintenance of real-world things collaborate with data scientists and IT professionals. Digital twins of physical assets combined with digital representations of facilities and environments as well as people, businesses and processes will enable an increasingly detailed digital representation of the real world for simulation, analysis and control." (Panetta, 2016).

Simulation approaches, which are using realistic models like the digital twin, are not new. In the aircraft industry the so called "iron birds" are used as a ground-based engineering tool when there is no prototype available. An iron bird is a ground based model in form of a building or a metal frame that includes all components of an airplane. By using the iron bird, first functionality and the interaction of the components can be tested. Nowadays digital versions of the iron bird are in common use (Airbus Group SE, 2017).

Within the smart factory an optimal image of the real factory is needed to enable a completely self-organized production and order-controlled factory. In this case, a cost-effective production of the lot size one would be possible. A flexible and autonomous system based on digital twins is, however, not yet available.

3.1 Overview of Digital Twins

The following section summarizes the use of state of the art digital twin technology from the perspective of the literature, as well as companies.

In the literature, digital twins can be found in different areas. Schuh et al. (2016) are describing an approach, which is used for order processing. According to them, it is necessary to have access to all the data related to the process. However, they point out that the data quality is not as important as is integrity.

According to Panetta (2016), Gartner foresees the use of digital twins in the manufacturing area, where they can be used for simulation, analysis and management of factories. Gartner's approach will require the implementation of sensors and a high performance software solution.

The Fraunhofer IPK (Institute for Production Systems and Design Technology) shares Gartner's view of digital twin implementation. Their research focuses on the management of the entire manufacturing process via a digital twin. They envision a bidirectional system, which allows direct access to manufacturing at any time (Stroheimer, 2016).

Södeberg et al. (2017) are seeing the digital twin concept in the area of real-time geometry assurance in individualized production areas. They describe a model in which the physical system is rebuilt in the digital world. Bidirectional communication between the physical- and the digital world is then used to continuously monitor and improve the production process. Access to large amounts of data, as well as the ability to analyze and understand those (which can be generated by different systems and therefore be heterogeneous), is important in all of these approaches. High performance software solutions are therefore crucial to the success of digital twins.

A practical example of the digital twin implementation is General Electric who have their own software solution called Predix. This high performance software solution is used for the building of digital twins of their assets (like windfarms, oilrigs, gas power plants, etc.). Predix is able to analyze and understand the data gained from sensors in the assets. The data collected can then be used for the status detection of the asset. Consequently, this approach allows a decrease in downtimes and the optimized planning of maintenance (General Electric, 2016).

Siemens also provides a high performance software solution called MindSphere for the realization of digital twins. MindSphere allows the connection of assets and the bidirectional communication thereof. The software can recognize important key performance indicators, such as downtimes and energy consumption, which are then relayed to the digital twin. This enables the creation of an accurate data model of the product and the production area, which leads to continuous improvement.

Bausch + Ströbel, a special-purpose machine manufacturer from Germany, uses MindSphere for the generation of a consistent engineering. Until 2020, this will lead to an efficiency increase of 30% in the process. Schunk, the world market leader in clamping technology implemented MindSphere in its engineering process for electrically-controlled gripping system. They are confident, that this extension of their engineering process will lead to significant decreases in project leadtimes, the commissioning process and the faster development of similar plants (Siemens, 2016).

The Austrian automotive manufacturer, Magna Steyr, is currently upgrading his machines to be able to generate digital images of themselves. The German mechanical engineering company Optima uses the digital twin in combination with a transportation system for simulations of operation setups and the production flow. They also use a simulation software offered by Siemens to build virtual replicas of their packaging machines for testing and validation (Hannover Messe, 2016).

The examples above show, that digital twins are continuing to evolve. With the decrease in costs for the basic technologies and their steadily increasing availability, digital twins will become more applicable for the factories of today. Identification, communication, localization technologies and high performance software solutions are building the core competencies for the digital twin.

Aspect/Paper	[1]	[2]	[3]	[4]	[5]	[6]	[7]	[8]	[9]	[10]	[11]
Connectivity	+	+	+	+		+		+	+	+	+
Availability											
Security	+		+			+					+
Speed	+	+	+	+	+	+	+	+	+	+	+
Cyber Reality	+	+	+	+	+	+	+	+		+	+
Flexibility and Variability	+	+	+	+			+	+	+	+	+
Globality											
Time											
Visualization	+		+	+	+			+	+	+	
Acceptance											

Table 1: Smart Factory requirements met by the Digital Twin. [1] Parrott and Warshaw (2017); [2] Alam and El Saddik (2015); [3] General Electric (2016); [4] Grieves (2015); [5] Schuh et al. (2016); [6] Kuhn (2017); [7] Rosen et al. (2016), [8] Schroeder et al. (2016); [9] Söderberg et al. (2017); [10] Stark (2017); [11] Uhlemann, Lehmann and Steinhilper (2017)

4 The evolution of the factory of today and the digital twin

As mentioned before, the factory of today needs to meet certain requirements in order to be called a smart factory and to meet the needs of the future of manufacturing. It is unlikely that this evolution will happen overnight. This paper already showed that more time is required for the concept of the smart factory to be adopted globally. However, digital twin concepts are ready and available for implementation.

Parott and Warshaw (2017) recommend starting with the implementation of the digital twin in a single area, to first deliver value there. Furthermore, once value has been seen in this area, implementation can expand throughout the factory until the digital twin has mirrored the whole system. This will enable companies to take a big step towards establishing a smart factory.

Evaluating the use of the digital twins for the smart factory implementation, 11 papers were examined in this paragraph. Table 1 tabulates which requirements of the smart factory are met by the digital twin and Figure 4 compares the requirements of the smart factory with the strengths of the digital twin as identified in the reviewed papers.

Furthermore, following the review of these papers, not all the requirements for the smart factory were discussed in light of the digital twin. The requirements of availability, security, acceptance, globality and the ability to work within different time zones are not specifically mentioned in the papers. Regarding to the requirements of acceptance, it can be assumed that successful implementations of digital twin in the initial area will lead to the acceptance of the concept in other areas. Successful implementation in the initial single area will therefore develop a positive attitude towards further implementation of the smart factory.

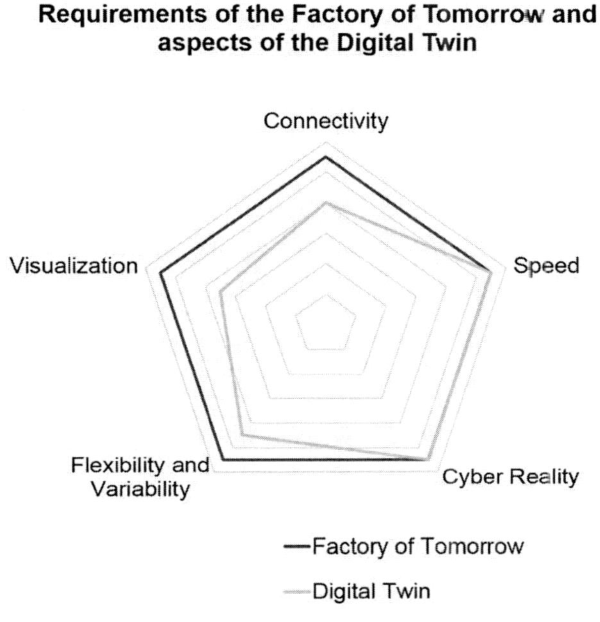

Figure 4: Comparison of the requirements of the Smart Factory with the strengths of the Digital Twin.

Due to the numerous smart factory requirements met by the digital twin, it can be concluded that the digital twin is a key technology for the evolution of the factory of today. Rosen et al. (2015) are also identifying the digital twin as an important aspect for the future of manufacturing in their paper.

5 Summary and Outlook

What is the link between the smart factory and the digital twin? This was the question to be evaluated by this paper. Therefore, both topics have been discussed. The outcome is that increasing variety of products and therefore the ever-increasing complexity of production processes are creating a situation in which new approaches are required. One new approach is the smart factory. Within this new approach, the digital twin is needed to enable the usage of its full potential. Some of the potentials are:

- increasing flexibility of factories and the production processes
- higher productivity,
- higher resource and energy efficiency,

- increase of transparency.

Thereby it is important to notice that the digital twin is actually used widely. For the industrial approach of the smart factory more research and development is needed. The technologies as well as their implementation and handling is still expensive. This might change due to fast development of new technology and therefore the ever-decreasing prices of those. Before the smart factory can be launched successfully, there is still more research and development required.

Literature

Airbus Group SE, editor (2017): Iron Bird. Retrieved from http://www.airbus.com/innovation/proven-concepts/in-design/iron-bird/ (accessed on Mai 3, 2017)

Alam, K. M. and El Saddik, A. (2015): C2PS: A Digital Twin Architecture Reference Model for the Cloud-Based Cyber-Physical Systems. IEEE Access, 5, pp. 205–206

Bundesministerium für Bildung und Forschung, editor (2017): Industrie 4.0. Retrieved from https://www.bmbf.de/de/zukunftsprojekt-industrie-4-0-848.html (accessed on May 01, 2017)

BOSSARD Proven Productivity, editor (2017): SmartFactory von Bossard. Retrieved from http://smartfactorylogistics.com/blueprint/images/blueprint-bg.png (accessed on Mai 10, 2017)

General Electric, editor (2016): The Digital Twin: Compressing Time to Value for Digital Industrial Companies.

Grieves, M. (2015): Digital Twin: Manufacturing Excellence through Virtual Factory Replication.

Hannover Messe, editor (2016): Klassische Fertigung auf der Schwelle zur Intelligenz. Retrieved from http://www.hannovermesse.de/de/news/klassische-fertigung-auf-der-schwelle-zur-intelligenz.xhtml (accessed on Mai 01. 2017)

Kagermann, H., Wahlster, W. and Helbing J. (2013): Umsetzungsempfehlung für das Zukunftsprojekt Industrie 4.0, Frankfurt/Main, acatech - Deutsche Akademie der Technikwissenschaften e.V.

Kellner, T. (2015): Wind In The Cloud? How The Digital Wind Farm Will Make Wind Power 20 Percent More Efficient. Retrieved from http://www.gereports.com/post/119300678660/wind-in-the-cloud-how-the-digital-wind-farm-will/ (accessed on Mai 09, 2017)

Kuhn, T. (2017): Autonome Systeme: Digitaler Zwilling, Fraunhofer-Institut für Experimentelles Software Engineering IESE, Kaiserslautern.

Lichtblau, K., Stich, V., Bertenrath, R., Blum, M., Bleider, M., Millack, A., Schmitt, K., Schmitz, E. and Schröder, M. (2013): 4.0-Readiness, Achen, Köln, Stiftung für den Maschinenbau, den Anlagenbau und die Informationstechnik.

Panetta, K. (2016): Gartner's Top 10 Strategic Technology Trends for 2017. Retrieved from http://www.gartner.com/smarterwithgartner/gartners-top-10-technology-trends-2017/ (accessed on April 15, 2017)

Parrott, A. and Warshaw, L. (2017): Industry 4.0 and the digital twin - Manufacturing meets its match, Deloitte University Press

Plattform Industire 4.0 (2017): Was ist Industrie 4.0?. Retrieved from http://www.plattform-i40.de/I40/Navigation/DE/Industrie40/WasIndustrie40/was-ist-industrie-40.html (accessed on October, 30, 2017)

Rosen R., von Wichert, G., Lo G., and Bettenhausen, K. D. (2015). About the importance of autonomy and digital twins for the future of manufacturig. IFAC-PapersOnLine, 48(3), pp. 567-572.

Scalabre, O. (2016): The next manufacturing revolution is here, TED Talks, Vancouver. Retrieved from https://www.youtube.com/watch?v=AyAyWtIwwE0\&t=44s (accessed on March 20, 2017)

Schließman, A. (2015): Zehn Anforderungen an weltweit eingesetzte Manufacturing-Systeme, Ravensburg.

Schroeder, G. N., Steinmetz, C., Pereira, C. E. and Espindola, D. B. (2016): Digital Twin Data Modeling with AutomationML and a Communication Methodology for Data Exchange, IFAC-PapersOnLine 49(30), pp. 12-17.

Schuh, G., Blum, M., Reschke, J. and BIrkmeier, M. (2016): Der Digitale Schatten in der Auftragsabwicklung, Zeitschrift für wirtschaftlichen Fachbetrieb, 11 (1/2), pp 48-51.

Shafto, M. Chair, Conroy, M., Doyle, R., Glaessgen, E., Kemp, C., LeMoigne, J. and Wang, L. (2010): NASA Technology Roadmap. DRAFT Modeling, Simulation, Information Technology & Processing Roadmap Technology Area 11, National Aeronautics and Space Administration.

Shafto, M. Chair, Conroy, M., Doyle, R., Glaessgen, E., Kemp, C., LeMoigne, J. and Wang, L. (2012): NASA Technology Roadma Modeling, Simulation, Information Technology & Processing Roadmap Technology Area 11, National Aeronautics and Space Administration.

Sherman, E. (2017): Is Slow Productivity Growth The Problem, As Greenspan Says, Or Is It Selfishness?. Retrieved from https://www.forbes.com/sites/eriksherman/2017/02/20/greenspan-blames-productivity-growth-for-economic-ills-but-inequality-big-factor/\#2901b666e5a5 (accessed on May 04, 2017)

Siemens, editor (2017): Twins with potential. Retrieved from https://www.siemens.com/customer-magazine/en/home/industry/digitalization-in-machine-building/the-digital-twin.html (accessed on October 23, 2017)

Sirkin, H. L., Zinser M. and Rose, J. (2015): Why Advanced Manufacturing Will Boost Productivity. Retrieved from http://www.bcgperspectives.com/content/articles/lean_and_manufacturing_production_why_advanced_manufacturing_boost_productivity/ (accessed July 17. 2017)

Stark, R. (2017): SmartE Fabrik 4.0: Digitaler Zwilling, Fraunhofer-Institut für Produktionsanlagen und Konstruktionstechnik IPK, Berlin

Strohmeier, K. (2017) Industrie 4.0: Virtueller Zwilling steuert die Produktion, Fraunhofer-Gesellschaft, München.

Söderberg, R., Wärmefjord, K., Carlson, J. S. and Lindkvist, L. (2017): Toward a Digital Twin for real-time geometry assurance in individualized production, CIRP Annals - Manufacturing Technology.

Pfeiffer, S., Lee, H., Zirnig, C., and Suphan, A. (2016). Industrie 4.0 – Qualifizierung 2025. Studie im Auftrag des VDMA (Verband Deutscher Maschinen- und Anlagenbau). Frankfurt/Main: VDMA.

Uhlemann, T. H.-J., Lehmann, C. and Steinhilper, R. (2017): The Digital Twin: Realizing the Cyber-Physical Production System for Industry 4.0. Conference proceedings The 24th CIRP Conference on Life Cycle Engineering (Procedia CIRP 61, pp. 335-340). Japan

Vogel-Heuser, B., Bauernhansl, T. and ten Hompel, M. (2017): Handbuch Industrie 4.0: Allgemeine Grundlagen (2nd edition). Berlin, Heidelberg: Springer.

Zillmann, M. and Wilk, C. (2016): Smart Factory – Wie die Digitalisierung Fabriken verändert: Transformation von der Werkshalle bis zur Unternehmensleitung, Mindelheim, Lünendonk GmbH.

On the Selection of Machine Learning Models

Volker Reichenberger[a]

[a] *ESB Business School, Reutlingen University, D-72762 Reutlingen*

Abstract

The increased availability of data gives rise to the use of machine learning methods for purposes like forecasting or quality control in operations management. Practitioners who want to employ these methods are faced with the task of choosing from a large number of available methods. We give an overview of classification methods and available implementations and present considerations for choosing appropriate methods.

1 Rise of the machines

The promises of the powers of artificial intelligence have experienced several ups and downs since the inception of the field in the 1950s. We are currently observing the strongest peak to date, mostly fuelled by successes in the field of neural networks, where e.g. speech recognition and face recognition made notable advances that can be experienced using a modern smart phone.

In this paper we focus on the task of binary classification, i.e. the task of deciding whether a given data input falls into one of two categories, or more specifically computing the probability with which the input falls into each of these categories. We present this in the framework of supervised learning algorithms, where a set of training data is available in which each input has been labelled with one of the categories. An example for applying this in operations management is the prediction of machine failures from sensor data: predict the probability of a machine failure based on a history of sensor measurements together with failure observations.

There exists an overwhelmingly large amount of algorithms for solving these kinds of problems (for an overview see Hastie et al.; Kuhn and Johnson (2009; 2013)). We seek to show some of the criteria that can be used when choosing appropriate algorithms. Apart from predictive power, we focus on the question of whether the model can be interpreted and on the algorithmic complexity. The former is important in order to be able to understand limitations of the model and assess the applicability to different situations. The latter is of special importance when large quantities of data are to be processed (*big data*). Some machine learning algorithms require a large amount of computing time for determining the model and these requirements can quickly rise to a level which makes it impossible to compute the model on the full training data, unless advanced concepts like massive parallelization are employed. For this paper we assume that access to a supercomputer is not available.

For reasons of both interpretability and runtime behaviour we exclude nonlinear learning methods like neural networks from our considerations. They are difficult or impossible to interpret and training can pose some serious challenges (e.g. overfitting).

When dealing with real world data some problems frequently arise (Hastie et al. 2009).

- Messy data: inputs consist of quantitative, categorial and binary variables, with possibly a large number of categories. Input variables may be measured on a wide range of scales.
- A large number of measured variables with many missing values, making complete sets of observations rare.
- A large number of inputs, which requires efficient methods from a computational perspective.
- Long-tailed and skewed variables (predictor and response) with outliers.

We comment on how the respective machine learning methods perform for these kinds of problems in the presentation below.

2 A Selection of Machine Learning Algorithms

There is a large number of machine learning algorithms that has been developed over the past decades. We focus on a selection of some popular algorithms which draw their popularity from their general effectiveness. However, this selection is very incomplete and we don't mention e.g. k-nearest neighbours, Bayes-methods or any type of neural networks.

2.1 Regression Methods

We are going to deal with the problem of making a prediction Y from an m-dimensional input variable X. Using observed data this can be stated as: *Given* n *observed values* $x_i = (x_{i1}, \dots, x_{im})$ *and* n *observed response values* $y_1, \dots, y_n$ *with* $y_i \in \{0, 1\}$ *for all* $i = 1, \dots, n$ *find a model which makes a good prediction of* Y *from* X. The observed values x_{ij} can be quantitative or categorical values.

A simple method for this kind of problem is logistic regression, which ensures that the output is between 0 and 1. Instead of the linear regression model function $Y = \hat{\beta}X$ the logistics regression model function is $Y = \exp(\hat{\beta}X)/(1 + \exp(\hat{\beta}X)$. While this method is seldom thought of as a machine learning algorithm, it fits into the

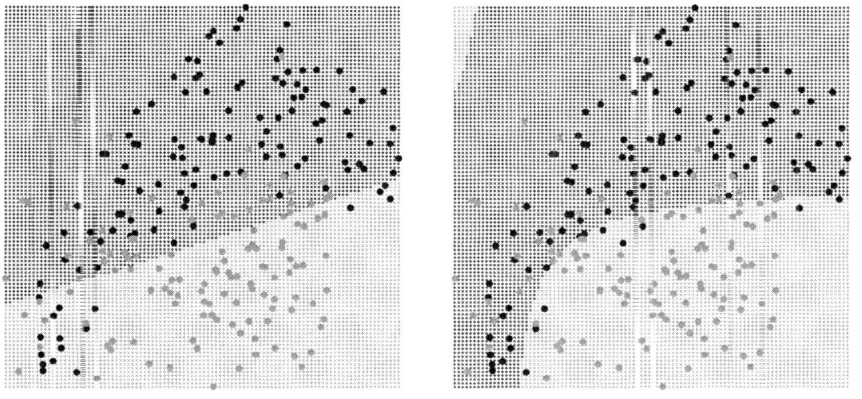

Figure 1: A sample classification problem and the results from logistic regression using a linear (left) model and a nonlinear model with mixed terms (right). The classification from the logistic regression model are the dots on the regular grid in the background, coloured depending on whether the predicted value was > 0.5 or ≤ 0.5

framework of supervised learning and has the large advantage that the theory is well understood (see Fahrmeir et al. (2009) for a complete discussion).

Among the logistic regressions advantages are its straightforward interpretation and that the model can be very efficiently computed. However, the regression puts several assumptions on the data, which are often not fulfilled in practise, and multicollinearity of input variables can pose a problem.

2.2 Trees and Forests

Tree-based represent simple, yet powerful models which partition the data into rectangular areas. There exists several variants of this method, the most prominent ones being the *Classification and Regression Tree* (*CART*) (Breiman et al. 1984) and *C4.5* (Quinlan 1993). The basic idea behind decision trees is to partition the data into more homogeneous subgroups. We start by dividing the full sample into two groups by searching an optimal split point for all dimensions of the data and then divide the data into two groups. The the subsets can be divided using the same methodology and this can be repeated iteratively. Different methods for finding the split points lead to different variants of the tree algorithm. Figure 2 shows the subsets for an example.

One of the advantages of tree methods is that the result can be interpreted very easily. Computing the decision tree is also very efficient.

One drawback of decision trees is that it is not trivial to decide at which level of refinement the method should stop. Overfitting poses an important problem here: By using as many refinement levels as input data a predictor results which classifies each

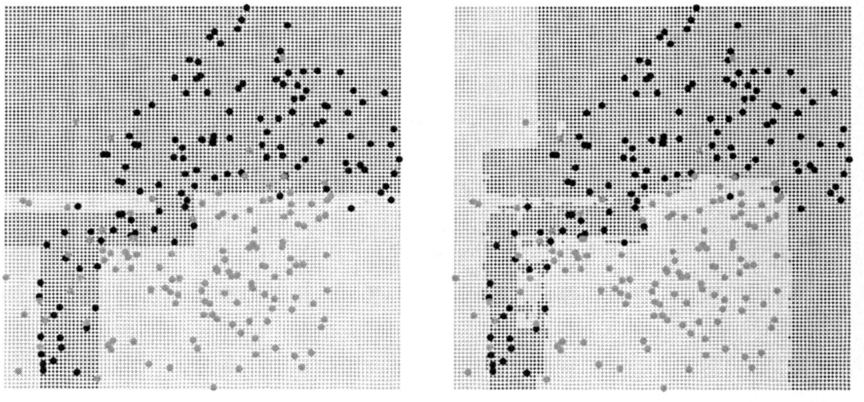

Figure 2: Classification using CART (left) and random forest (right).

input exactly, but which is not well-suited for non-learned input.

Fortunately decision trees can be improved by two simple ideas: *bagging* computes not one but many decision trees based on different random samples of the data. Each of these trees is different and the classification is done by averaging the votes of each trees. In a very successful method called *random forests* (Breiman 2001) this is extended further by not only using different sub-samples but also by using random sub-sets of the input variables. An example for a random forest classification is shown in figure 2.

2.3 Boosting

The motivation behind boosting is that the combination of many so-called *weak classifiers* or *weak learners* can lead to a committee with much increased predicitve power. One of the most successful boosting algorithms is *AdaBoost.M1* (for *adaptive boosting*, see Freund and Schapire (1997)), which has often been labelled the most powerful out-of-the-box classifier. Several variants of boosting algorithms exist.

AdaBoost assigns a weight to each input vector in the training set ($1/n$ in the beginning). We assume that there is a weak classifier which can be trained on this weighted data and produce a prediction. Then the algorithm changes the weights in each iteration so that misclassified points receive more attention. When the desired number of classifiers has been trained, predictions can be made by a committee decision from the weak classifiers using different weights for each weak classifier. The described procedure enables the algorithm to focus on the difficult-to-learn features of the data. Because AdaBoost concentrates on the features which provide the largest improvement to the model, the algorithm effectively reduces the dimensionality of the problem and thus the execution time. An example for an AdaBoost.M1 classification classification is shown in figure 3.

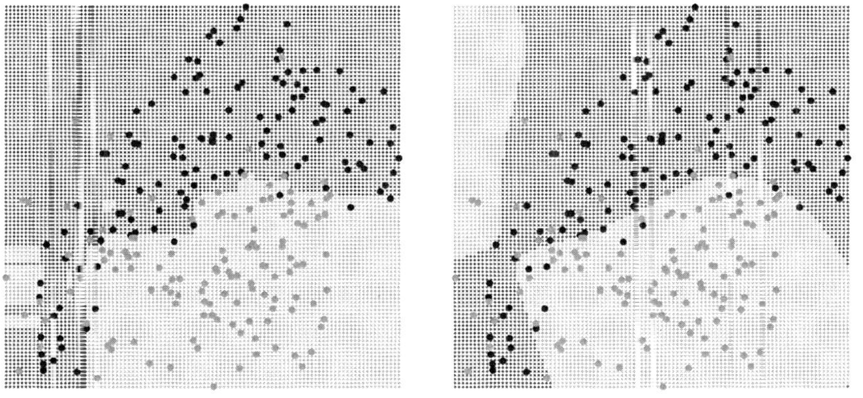

Figure 3: Classification using AdaBoost and support vector machine.

2.4 Support Vector Machines

Support vector machines (SVM, Vapnik (2010)) are based on maximum margin classifiers: If a data set consists of two subsets which are completely separable by a linear boundary (for two-dimensional data this is a line which separates the two sets), then there are in general infinitely many linear boundaries which separate the subsets. Since all of these boundaries completely separate the subsets, they are all equivalent with respect to performance measures (accuracy). Maximum margin classifiers choose the boundary for which the distance between the boundary and the closest points is as large as possible. This approach can be extended to non-separable data and to to nonlinear boundaries based on *kernel functions*. Figure 3 shows a nonlinear boundary computed with a SVM.

Support vector machines are popular because the show a large predictive power in many situations. They can also be employed to extract linear combinations of features. They are, on the other hand, notoriously difficult to interpret.

3 Model Selection

3.1 Considerations for Model Selection

Choosing an appropriate model is a difficult task and the performance of different algorithms is highly dependent on the features of the data and the model response variable. Below we present a selection of considerations that should be taken into account.

- The first one is that the model should have *predictive power*. Several metrics are available for assessing the predictive power of a model–e.g. the receiver operating characteristic (ROC)—but the correct choice of a metric depends on the underlying problem. Models with higher predictive power are preferable over models with lower predictive power, but if several measures are used the results can contradict themselves.

- The second consideration for model selection is that the result should be *interpretable*. The difference between data mining and pattern recognition is that in data mining a qualitative understanding of the construction of the predicted values from the predictor variables is important. This also implies that simpler models should be preferable over more complicated ones, so that e.g. a linear regression model should be preferred over a nonlinear regression model if they have similar predictive power.
- A third consideration is the *scalability* of the algorithm for large samples. Some algorithms have an algorithmic complexity which would require prohibitively long computing times for large data sets. Of course, this feature depends to a very large extend on the implementation of the algorithm–below is an example for two implementations of the support vector model with very different run time behaviour.
- Properties of the data may make it difficult or impossible to employ certain methods. The data may exhibit mixed type values (quantitative and categorial), outliers or missing values.

When applying these considerations to the methods presented above, the following picture arises.

Regression methods can achieve high predictive power, at the cost of having to choose the appropriate set of variables. Regression models may require some serious data preparation, because the method may not work well on the raw data. Regression models can be computed very efficiently. Regression can handle mixed data types, but is vulnerable to outliers and missing values. One of the largest advantages of regression models is their straightforward interpretability.

Tree methods come close to out-of-the-box methods which can be applied in most circumstances. They are inexpensive to construct and easy to interpret. Mixed types, outliers and missing values do not pose a problem. The drawback of single tree models is often the predictive power, which can be enhanced greatly by random forests–at the cost of losing interpretability and scalabiltiy.

Boosting can improve the predictive power, sacrificing speed and interpretability (and in the case of AdaBoost robustness against overlapping class distributions and mislabelled training data).

Support vector machines have very high predictive power, but they rate poorly on scalability, robustness to outliers, and the handling of mixed data types and missing values.

Now given the task of finding an appropriate model for real life data, we propose to proceed as follows.

1. Start with methods which promise large predictive power in many cicrumstances. AdaBoost, support vector machines and random forests can be applied in many situations and can create highly accurate results without much manual tweaking.

2. Proceed to simpler models which are easier to interpret. Variants of (logistic) regression models can be very accurate given the right choice variables (this might include data transformations which leads to more suitable input variables). Maybe tree models can compete with the more sophisticated competitors.

3. If possible, choose one of the simpler models which has a predictive power that is comparable to the more complex models.

3.2 Implementations

Many software systems are available for using machine learning algorithms. Some prominent examples are RapidMiner (Hofmann and Klinkenberg 2013), WEKA (Hall et al. 2009) or systems based on the R language and environment (R Core Team 2017). We used the latter for running some computer experiments to compare the scalability of some popular methods. Building models in R is usually fairly simple, since different libraries implement the aforementioned models. Given a vector of responses `y`, a matrix of input values `m` and a so-called data frame (mostly a table) `df` which combines `y` and `m`, a linear regression model, a random forest, AdaBoost and a support vector machine can be created in this order as follows.

```
glm(z ~ ., family=binomial(link='logit'), data=df);
library(randomForest);    randomForest(m, y=zf)
library(ada);             ada(x=m, y=zf, type="discrete")
library(kernlab);         ksvm(z ~ ., data = df, type="C-svc")
```

We created these models on different data samples, with data sample sizes from $n = 512$ to $n = 65024$ on a Intel Xeon X7550 2.00GHz system with 4 GB RAM. It has not escaped our attention that this computer experiment compares the implementations rather than the methods. To show the importance of efficient implementations, we used two different support vector machine implementations, the aforementioned method from the `kernlab` library (ksvm) as well as the implementation from the `e1071` library (svm). The difference in algorithmic complexity is very notable in figure 4.

It is also obvious that linear regression can be computed very efficiently, while AdaBoost and random forest have close to linear scalability, but with much larger constants. The support vector amchine is very fast for small problems ($n < 5000$), but with increasing problem size AdaBoost performs faster (from $n > 10000$). The

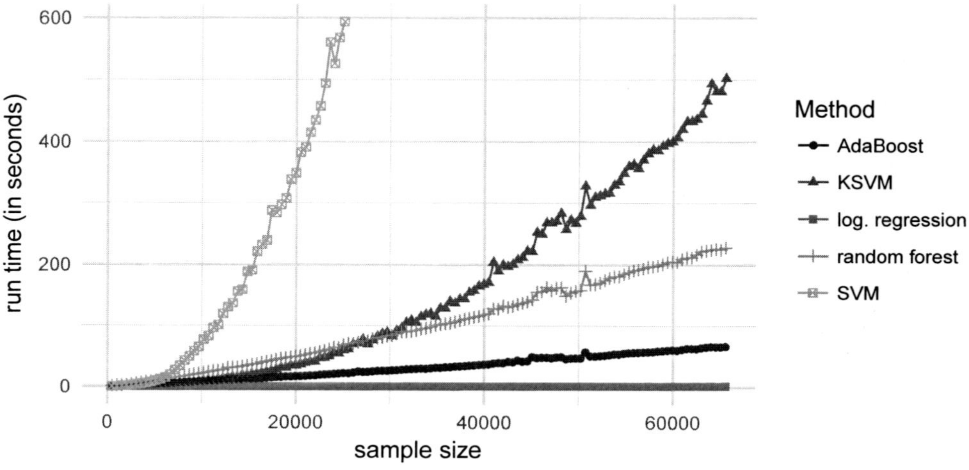

Figure 4: Run times of different machine learning agorithms.

computing time requirements of more machine learning algorithms have been studied in Lim et al. (2000).

References

Breiman, Leo (2001). "Random Forests". In: *Machine Learning* 45.

Breiman, Leo et al. (1984). *Classification and regression trees.* Monterey, CA: Wadsworth & Brooks/Cole Advanced Books & Software.

Fahrmeir, Ludwig, Thomas Kneib and Stefan Lang (2009). *Regression: Modelle, Methoden und Anwendungen.* Springer-Verlag.

Freund, Y. and R. Schapire (1997). "A decision-theoretic generalization of online learning and an application to boosting". In: *Journal of Computer and System Sciences* 55.

Hall, Mark et al. (2009). "The WEKA data mining software: an update". In: *SIGKDD Explorations* 11.1, pp. 10–18.

Hastie, Trevor, Robert Tibshirani and Jerome Friedman (2009). *The Elements of statistical learning. Data Mining, Inference, and Prediction.* 2nd ed. Springer.

Hofmann, M. and R. Klinkenberg (2013). *RapidMiner: Data Mining Use Cases and Business Analytics Applications.* Chapman & Hall/CRC Data Mining and Knowledge Discovery Series. CRC Press.

Kuhn, Max and Kjell Johnson (2013). *Applied Predictive Modeling.* New York: Springer-Verlag.

Lim, Tjen-Sien, Wei-Yin Loh and Yu-Shan Shih (2000). "A Comparison of Prediction Accuracy, Complexity, and Training Time of Thirty-three Old and New Classification Algorithms". In: *Machine Learning* 40.

Quinlan, J. R. (1993). *C4.5: Programs for Machine Learning*. San Mateo: Morgan Kaufmann Publishers.

R Core Team (2017). *R: A Language and Environment for Statistical Computing*. R Foundation for Statistical Computing. Vienna, Austria.

Vapnik, V. (2010). *The Nature of Statistical Learning Theory*. Springer.

Mitigating the risk for technology related investments in logistics: the STIC analysis as a decision support tool

Augusto Urru[a], Marco Bonini[a], Wolfgang Echelmeyer[a]

[a] *ESB Business School – Logistikfabrik Department, Reutlingen University, D-72762 Reutlingen*

Abstract

The success of an autonomous robotic system is influenced by several not easily identifiable interdependent factors. This paper is set to lay the foundation of an integrated approach in order to examine all the parameters and understand their contribution to success. After introducing the problem, two autonomous systems for the process of unloading of containers are presented. Then a recently developed method for modelling and interpreting all the parameters, the STIC analysis, are introduced. The preliminary results of applying such a methodology to a first study case, is shortly presented. Future research is in the end recommended in order to prove that this methodology is the only way to overcome barriers to the investment in autonomous systems in the logistics sector.

1 Introduction

One of the most ambitious research goals of the last few years was to make robotic systems autonomous. Thanks to the latest advances in software and hardware technologies, autonomous systems have reached in many cases a satisfactory level of development and are fairly ready to be introduced to the market. These autonomous or cyber physical systems are identified as one of the most relevant technologies and contributing concepts to the future development of the industrial revolution (Pfohl, Yahsi and Kurnaz, 2015). In the ranking made by Angelov (2012) the second distinguishing characteristic feature of the 4th industrial revolution is the autonomization: "Industry 4.0" technologies and concepts are enabling machines and algorithms of future companies to make decisions and perform learning activities autonomously. This autonomous decision-making and learning is based on man-made algorithms and enables whole factories and manufacturing facilities to work with minimum human machine interaction (Pfohl, Yahsi and Kurnaz, 2015). Even if logistics is considered since years one of the most promising market segments, innovation and new technologies struggle to penetrate the market. While automation became increasingly popular in logistics applications, robotics has been applied only thriftily in this field. Robotics for the automation of logistics processes usually presents higher costs in respect to the benefits and higher technical difficulty than in the field of manufacturing and autonomous robotic systems seem to be the missing piece of the puzzle that could enable a fully automated material handling from input to output of every logistic facility (Bonini et al., 2012). The challenge for both investors and technology providers is to bring new technologies to the market enabling the innovation through a technology driven process. Unfortunately the strategy of technology developers and system integrators often does not follow a standard procedure to analyze the target logistic environment and the available specific technology. As a consequence the gap between expectations and reality, requirements and technical specification, academy and industry is too wide to be filled and the new cutting-edge technology often does not reach the market.

The goal of this paper is to provide investors and researchers a structured procedure to support the process of fine-tuning and adaptation of both processes and technology in order to speed up the process of developing and bringing new autonomous technology to the market, filling the existing gap between technologies and applications.

First a brief description of the available autonomous robotic systems for container unloading will be given. In the following sections the method (STIC analysis) will be introduced and the preliminary results of the research presented. The discussion and forthcoming research chapter will bring the reader to the last section of conclusions.

2 Autonomous container unloading – State of the Art

In a recent study of the DHL group a robot operating in logistics is defined as follows: "A robot with one or more grippers to pick up and move items within a logistics operation such as warehouse, sorting center or last-mile" (Bonkenburg, 2016). Taking this definition of a robot operating in the logistics environment as a starting point one could notice that, despite the simplicity of the pick and place tasks, the variety of the application field is quite wide, ranging from the operational environment of a warehouse to the totally unconstrained and unpredictable environment of the last mile delivery. To date no tool exist that can help investors and researchers to bring a new technology to the market; this is mainly due to a lack of communication as each player tends to approach the problem from its own point of view. Therefore, both investors and researchers are unable to identify all the leading-to-success factors. Logistics is a complex and various environments in which procedures, people and business are often subjected to sudden changes: in such undisciplined environment is all but immediate to understand how different factors interact and influence each other. The development of a new autonomous robotic system for logistics is a delicate and ambitious challenge in which technology cannot be pushed to customers without its contextualization within the forthcoming market. Hence a sort of mutual collaboration between the academic (research) and the industrial world (real environment scenarios) is neded: on the one hand the technology needs to be developed, on the other hand the focus needs to be set on the final goal of bringing the innovation to the market, hence completing the innovation cycle. The authors of this paper have collected experience in the last 6 years in the field of autonomous robotics system for containers unloading, addressing both research and real-world scenarios. Thanks to the research work performed within the RobLog project (RobLog, 2011) and to the consulting activities in Europe and the USA, both academic and industrial world have been merged, taking into account requirements and peculiarities of a variety of logistic processes. Hereafter the two systems available to the authors for analysis and tests are shortly described.

2.1 The Parcel Robot PR 500

The autonomous container unloading is one of the still unsolve issues in logistics even if solutions have been developed, namely: (1) the Parcel Robot PR 500 (Parcel Robot PR 500), (2) TEUN's PIQR01 (TEUN's PIQR01) and (3) Wynright's RTU (Wynright, 2013). Since the TEUN system works only in fixed defined scenarios, while the Robotic Truck Unloader is still (since 2012) under development, the latest version of the Parcel Robot PR500 is currently the only ready-to-market fully autonomous robotic system for unloading of loosely loaded containers, trailers and swap bodies (Bonini et al., 2012). The design of the Parcel Robot technology began 13 years ago, in 2003. The first prototype was the result of a research project in which the goal was the development of a system for the Courier Express Parcel (CEP) branch able to cope with all the possible parcels in all the possible orientations that needed to be unloaded from a container. Flexibility was the key factor and the focus was on technology and effectiveness. The final prototype of the system was surprisingly effectively, but with an embarrassing performance of 230 parcels per hour, not even comparable with the average productivity of an operator. ThyssenKrupp carried the development of the machine on and in 2007 developed the PR400. This system still represents the most flexible and reliable Parcel Robot version. Even if the performance had been improved up to 450 parcels per hour, the 6 degrees of freedom (DOF) architecture combined with a heavy gripper were not enough to meet the customer requirement. In 2009 the last version of the Parcel Robot (PR 500) was developed. Less expensive thanks to the standard components employed and with a performance up to 1800 parcels per hour in case of sustainable triple-grasping, this version still is the most performing version of the system. In order to reach such a high performance the kinematics needed to be adapted and the machine had been downgraded from 6 to 4 DOF penalizing the flexibility of the system.

The story of the development of the Parcel Robot is a perfect example in order to understand the complexity and the ambitious challenges of logistics. Version after version, the evolution of the system can be analyzed as follows, pointing out the focus of each new version:

1. At the first development stage the system was able to perform the task but with low performance. It was not of interest for the industrial world;

Figure 1: Parcel Robot – PR 500

2. With the PR 400, the previous version was "industrialized". The focus was still on the technological aspect and some requirements coming from the industrial scenario such as performances (that had been improved) and reliability. But still the system was not ripe to reach the market mainly because of the high price for the logistic CEP branch it was developed for (needing for instance the 6 DOF expensive kinematic);
3. The development of the last version PR 500 took into account also the economic aspect as it was less expensive and technologically improved. The major barrier that stopped the system to reach the market depended on the scenario. The machine was not able to fulfill the requirements of the CEP branch anymore. The flexibility dropped down (due to the 4 DOF kinematik intead of 6) and so did the desirability. The Human-Robot interaction was not yet taken into consideration.

2.2 The RobLog Industrial Demonstrator

The Roblog industrial demonstrator was developed to meet the requirements of the coffee industry (RobLog industrial demonstrator). This autonomous system is able to autonomously unload coffee sacks from a floor loaded standard container. Two state-machines work together managing the decision the system is called to take: one is designed to handle the various motions of the robot, the other to handle and coordinate object recognition, object-selection and grasping configuration for the selected object (RobLog Consortium). This allows the system to identify and select the next best-candidate (sack) to be unloaded, to reach the right position in order to start the unloading process and finally to grasp and unload the sack; all of this under continuous and real-time monitoring of the workspace in order to avoid collision and promptly react to unforeseen events. Since the system is still in an early stage of development, the circumstances are not quite the same as for the PR 500. According to the definitions of the EU Horizon 2020 funding program the system after the end of the project has reached a Technology Readiness Level (TRL) of 6, namely a "technology demonstrated in relevant environment" (European Commission, 2014). In order to reach the market the technology needs to be prototyped, tested and optimized (TRL 7: "system prototype demonstration in operational environment"), completed and qualified (TRL 8) and finally proven in operational environment (TRL 9) (Urru et al., 2015).

3 The STIC Analysis

In the previous section the two systems available for deep analysis to the authors have been described, in the following the method will be explained. The methodology developed consists of a well-balanced mix between strategical and technical aspects that are to be considered in order to bring an autonomous system to a skeptic market such as logistics. When it comes to complex and innovative robotic systems, the customer is often diffident and not willing to invest (Khan, 1993). Therefore every detail needs to be carefully considered especially when a technology has reached the highest level of Technology Readiness Level (TRL) but it still struggles to reach the market, becoming a real successful product. Scenario related, Technological, human-robot Interaction and economic Convenience aspects need to be not only investigated but analyzed together in order to deeply understand their nature and the delicate reactions that even the smallest change in one of them could trigger, with potential impact on the sought final result. Moreover the unpredictability of the logistic sector leaves no room for mistakes. The STIC (Scenario, Technology, Interaction and Convenience) analysis explores systematically every aspect and combines all the objective metrics in a model for the calculation. It also gives an approach for the interpretation of the results in order to suggest how to steer the introduction of the new technology into the adverse logistics market. Hereafter the 4 pillars of the STIC analysis and the subsequent integrated approach that enables the interpretation of the results are described.

3.1 The four aspects of the STIC analysis

3.1.1 Scenario

The scenario analysis aims at pointing out all the on-site aspects of each specific study case or innovation project. On the one hand, the physical environment, on the other hand the peculiarity of the business and the current processes. The first step is often an on-site visit where all the key features are collected and the requirements elicited. Once the scope of the automation project has been defined, the first constraint to deal with is the physical one, namely the available space in the current layout. Right after a detailed process analysis aiming at identifying interfaces and collecting data regarding bottlenecks, current performances and standardization level of the selected process needs to be performed.

3.1.2 Technology

After the scenario has been investigated, the system needs to be analyzed from the technological point of view. Firstly, all the components need to be listed and singularly analyzed in order to identify the relation between the components and the overall performance of the system. It is good practice to reconsider critically the system components. It is often the case that, before reaching the market, products went through a long development process that could have lasted years. It might happen that some of the hardware (sensors, grippers, conveyors) or software components are not the latest version available on the market. If this is the case not only the replacement of these components could increase the efficiency and decrease the costs of the system itself, but also the simple preventive maintenance could take some advantages out of this upgrade. Other parameters such as sustainability, reliability and precision, essential in the industrial field, could also benefit. Besides the research work that needs to be carried out in order to investigate alternative, new and better technical solutions, the integration effort must not be underestimated. Industrial sensors and actuators are mostly not plug and play devices and the system might need to be further developed or slightly adapted with all the engineering-related work, let alone the potential chain reaction that one small but necessary change in the system can create with impact on other modules of the system.

3.1.3 Interaction

Logistics has been defined by the Canadian Association of Logistics Management as "the process of planning, implementing, and controlling the efficient, cost effective flow and storage of raw materials, in-process inventory, finished goods and related information from point of origin to point of consumption for the purpose of meeting customer requirements" (Canadian Association of Logistics Management, 1998). Starting from this definition it is clear that both material and information flows have to be handled together. Unlike in production, standardization is not logistic's strong suit; moreover the point of origin and the point of consumption of a good (material flow), as well as the point of acquisition and the point of use of an information (information flow), can be distant in space and time, which creates opportunities for unforeseeable events to occur. As a result, autonomous systems alone might not be robust enough in logistics to ensure a reliable process in case of unforeseeable events. A prossible solution for this problem is the inclusion of the human in the process chain, enebling Human-Robot interaction through a smart interface (Bonini et al., 2012). Evaluating the efficiency in Human Robot interaction is not an easy task and often requires tests and a detailed Methods-Time Measurement (MTM) analysis (Maynard, Stegemerten, and Lowry, 1948). One of the most important quantifiable indicators in human-robot interaction and collaboration is the fluency. According to the objective method for evaluating the level of fluency proposed by Hoffman (2013), not only the fluency has a non-negligible impact on the overall efficiency of the system, but also it positively influences the acceptance of the new robotic system by the workers.

3.1.4 Convenience

In order to penetrate the market with a new technology the efficacy of the autonomous system is not the only enabling criteria. The investment in the new technology needs to be economically convenient. Even if basic logistic tasks generally do not change from business to business, the operational environment and consequently the business framework often do. It is essential to target the most suitable logistic branch for the system to be efficient and so convenient. The three main branches Courier Express Parcel (CEP), contract logistics and production logistics have their own specific characteristics which could enable or defuse the innovation process, making the investment attractive and worthy (Khan, 1993). The model used to evaluate the economic convenience of an investment is usually the well-known Life Cycle Cost Analysis (LCCA), but for investors in the intralogistics world the main parameter considered is the Net Present Value (NPV) (Khan, 1993) function with special regard to the Discounted Payback Period (DPBP) (Bhandari, 1989). According to the experience collected on the field, the main concern of companies is the time the investment will pay back. When considering innovation in logistics this is not a fixed defined point in time but it is more a time window usually set between 2 and 3 years after the investment due to the unpredictability of the business and the high risk perceived. It can vary from business to business but there is a trend in each logistic branch. In the case of production logistics, this threshold can be extended up to 4 years (Bononi et al., 2015). In this economic phase of the analysis, all the economic and financial factors of the target company useful in order to calculate the NPV function needs to be identified and quantified. Country and regulations can play an important role in determining the values of these factors (e.g. inflation, tax shield etc..).

3.2 The Integrated Approach

Once all the aspects of the STIC analysis have been taken into account, the most difficult task is to quantify the influence they have on each other and to understand to what extent a specific change in the one or the other direction could positively impact the overall success to the market. The main challenge the system producers or developers need to face is to identify the weaknesses of the new system and give them priority; this helps focusing the improvement process on the right aspect. Given the complexity of both autonomous systems and logistics environment the relations between parameters and metrics are not often clearly identifiable and the magnitude of the effect one change could cause on the overall system is not easily foreseeable. For these reasons an integrated approach needs to be considered in order to correctly channel the efforts in the fine-tuning of the autonomous system. As mentioned in section 3.1.4, the main driver for companies evaluating the possibility of investing in a new technology is the Discounted Payback Period (DPBP). For this reason, once each aspect of the analysis has been taken into account, it is possible to extrapolate a set of parameters that will be the input for the integrated approach. On the basis of the parametric model developed by Bonini et al. (2012) a new version of the tool is currently under development in order to include in the input parameter sheet all the useful informa-

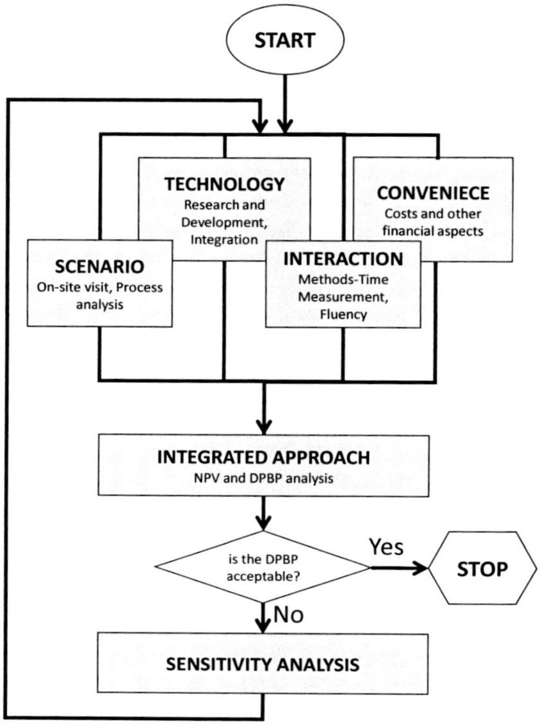

Figure 2: The STIC analysis method scheme

tion gained through the STIC analysis. Thanks to its parametric nature the new tool can easily describe and analyze every logistic process. Last but not least a sensitivity analysis needs to be performed in order to identify the ranking of the most influencing parameters on the DPBP. The STIC analysis method scheme is shown in Figure 2.

The STIC analysis not only guides the analyst throughout the data gathering process and the identification of the most sensitive parameters; it also provides guidance for their quantification, so that they can be numerically given in input to the calculation model obtaining the NPV function and the DPBP. If the result is not satisfactory, a sensitivity analysis must be conducted in order to identify the most influencing factor and put in place the corrective actions. After the improvement of the most influencing factors, the analysis could be repeated in order to assess the new balance among the different aspects. Thanks to this integrated approach, it is possible to consider both the point of view of the investor and the point of view of the developer or system integrator, without forgetting the analysis of the framework and the final user (worker). This provides the right balance for the development of a competitive autonomous robotic system able to penetrate an adverse market such as logistic.

4 Preliminary results

The core competence of the research group atoring this paper lies in autonomous systems for container unloading. The STIC analysis was first applied to the latest version of the parcel robot PR 500 and the preliminary results are hereafter shortly presented:

The scenario analysis started with an on-site visit at a facility where the customer needed a solution to unload containers loosely loaded with parcels. All the process related aspects and the characteristics of the facility have been analyzed and the parameters collected and listed. The analysis made it possible to feed the right input parameters to the developed model such as process performance, both in peak period and average, average trailer set-up time, working hours per shift, layout constraints, etc. Then specific parameters and indicators technology related have been identified. Every single activity of the autonomous system was deployed into simple tasks until relations between system performance and system components have been deeply analyzed. Scanning, grasping, unloading and positioning processes, failure acknowledgment time, workspace of the robot in working position, etc. From the interaction analysis it turned out that the robot had been designed to be 100% autonomous. The interaction with the human was not foreseen. As a result the efficiency of the system dropped down each time the human intervention was needed, hence in all those unforeseable cases the autonomy was not able to solve. Through an MTM analysis the process of human intervention was analyzed and quantifiable parameters were obtained, such as trailer access time, failure aknowledgemnt time for the operator, time for restarting the robot, etc.; In the specific case of the PR500, two scenarios have been taken into account, one considering small parcels, the other considering big parcels. As shown in Fig. 3, the impact of one damaged parcel, which requires human intervention, has a non-negligible impact on the overall autonomous system performance.

The economic analysis contributed to identifying all the economic and financial parameters of the target company useful in the numerical calculation of the NPV function and the DPBP. After analyzing all the aspects both from the scenario and the technology point of view it was possible to feed the right input to the calculation model and to perform the sensitivity analysis, based on the NPV function optimization, in order to prioritize the possible improvements. The ranking of the most influencing parameters has been calculated and urgent improvements have been identified. Combining the results of the sensitivity analysis with the data collected in the scenario, technology and interaction analysis room for improvements have been identified especially concerning sensors and the conveying technology that needed to be slightly modified in order to meet the process requirements. The technology was relatively old in comparison with the speed of innovation in this field, indeed. This analysis has also identified some lacks in the gripping technology that have been easily fixed, improving the reliability of the system. On the other hand, also some points of weakness have been identified in the logistic processes at the specific facility. Current processes were partly reengineered in order to fully exploit the potential of the autonomous sys-

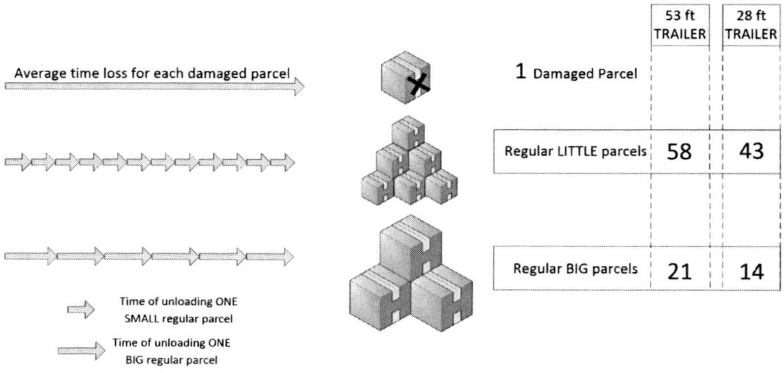

Figure 3: Average time loss (in seconds) due to human-robot interaction

tem. Thanks to the integrated approach, it was possible to clearly isolate mistakes in the human-robot interaction activities. The Interaction analysis helped identifying the restricting factors and improving the fluency of the operations: this happened simply through a smarter division of tasks between robot and human. Furthermore, the economic analysis has contributed to identify the specific logistic branch in which this kind of technology is more suitable, namely the production logistics (Canadian Association of Logistics Management, 1998), which, thanks to its nature, has been selected as the closest market to be reached by this kind of innovation. The new version of the PR 500 is being produced by a German robot producer and the new system will be tested by 2018 at a customer's facility. After another round of STIC analysis minor adaptation are still necessary before the system could reach the logistic market, but preliminary results are encouraging. After the set of input data for the calculation has been updated, the DPBP was reduced by more than three years and the system is now less sensitive the lack of standardization, which characterize the logistic sector (Fig. 4). The risk of investing in this autonomous system has been therefore mitigated and the skepticism of potential investor reduced.

5 Discussion and forthcoming research

One of the strong statements of the analysis is that the approach needs to be integrated. This means all the aspects needs to be considered together in order to guide the improvement process and finally lead the innovation to the market. The evidence that a disjoint approach is not successful is the Parcel Robot development history itself. In the first stage of development the focus was set only on the scenario and the technological aspects. The result was an expensive flexible system able to cope with almost all the possible unloading scenarios, but with poor performances. The second version of the Parcel Robot, the PR400, was more industrial then the previous one,

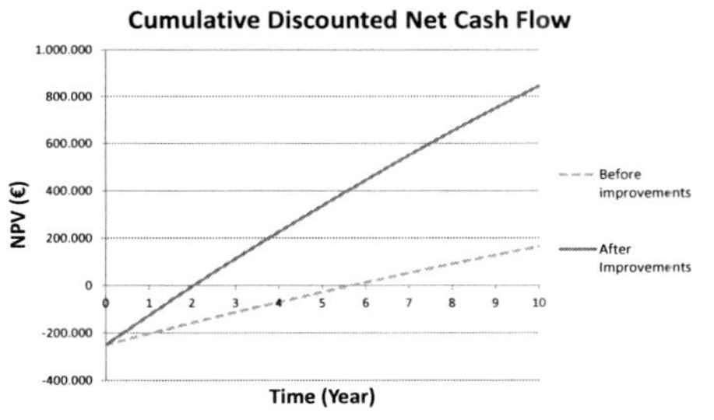

Figure 4: Cumulative Discounted Net Cash Flow

but the focus of the developers was this time mainly on the technology; as a result the PR400 was reliable, flexible and more performing than its predecessor, but the processes requirements were not fully considered. This was especially true for those requirements coming from the CEP branch: even if the performances of the system had been doubled the growth of the e-commerce business and the evolution of the sorting systems made the requirements of the CEP provider stricter than ever. This forced the developers to reconsider the economic aspect in order to produce a less expensive and more performing system. Improving performances and reducing costs came not without a cost: the flexibility of the system was drastically reduced, mainly due to the reduction of the degrees of freedom of the machine from 6 to 4. This important decision led to two main consequences: first, the CEP branch, where the standardization level is the lowest in logistics, lost interest in the system; secondly the fact that the human-robot interaction was not foreseen played a crucial role in the overall system performances and acceptance. The elaborate and time-consuming procedure to solve the system failures made both investors and operators skeptical. In retrospective, these are the main reasons, identified and quantified thanks to the STIC analysis, why also the PR500 was not successful on the market. Despite expectations, the perfect system able to cope autonomously with 100% of the possible unforeseeable situation that can occur in logistics in an efficient way has not been designed yet. However, in order to reach the market a product does not need to be perfectly autonomous. What is needed and requested by the market today is a compromise, that gives investors and workers enough guarantees concerning reliability, performances and a high level of acceptance. Only through an integrated approach that considers the four aspects of the STIC analysis, the most influencing factors could be identified and the effort could be focused in improving the most relevant one; this leads to the fine-tuning and adaptation of the autonomous robotic system making the technology attractive for the market. The STIC analysis provides therefore a risk-management tool that can reduce skepticism of investors and users of new technologies.

However, further research is needed to optimize the calculation model and to prove that the method is generally applicable to other autonomous technologies operating in logistics environment. The following steps are foreseen to continue the research on this topic:

1. Fully implement the fluency factor in the calculation model. The fluency factor is currently only partially considered due to the specific nature of the first study case;
2. Apply the STIC analysis to a less developed technology with a TRL lower than 9, such as the RobLog Industrial Demonstrator;
3. Apply the STIC analysis to a technology that is not designed for the unique process of unloading. The research group is involved in a project for the design and development of an intelligent turning unit that could be developed exploiting the STIC analysis and this is encouraging for the future.

6 Conclusions

In this paper were first introduced challenges and obstacles that both investors and technology providers need to face in order to successfully bring a new innovative technology to the market. After a brief review of two autonomous systems for the unloading of containers, a method for modelling and interpreting all the parameters has been proposed. The STIC analysis was implemented and the potentiality of the method was shown through a real study case and supported by preliminary results. However, future research is needed in order to optimize the model and prove the strong statement that the STIC analysis could successfully guide the optimization and the fine-tuning of a generic autonomous robotic system assuring its success in an adverse market such as logistics.

Literature

H.C. Pfohl, B.Yahsi and T. Kurnaz "The impact of Industry 4.0 on the Supply Chain", Hamburg International Conference of Logisitc (HICL) 2015

P. Angelov "Autonomous Learning Systems: From Data Streams to Knowledge in Real-time", John Wiley & Sons, 2012

M. Bonini et al. "Towards the full automation of distribution centers" in proceedings of 4th International Conference on Advanced Logistics and Transport (ICALT) 2015, pp.47-52, 20-22 May 2015 doi: 10.1109/ICAdLT.2015.7136589

T. Bonkenburg "Robotics in Logistics – a DPDHL perspective on implications and use for the logistic industry", DHL Customer solution & Innovation, Deutsche Post DHL Group, March 2016

RobLog - Cognitive Robot for Automation of Logistic Processes – Grant Agreement FP7 – 270350, www.roblog.eu, 2011

Parcel Robot PR 500 http://www.cemat.de/product/parcel-robot/487286/S731673

TEUN's PIQR01 information. Available online: http://www.teun.com/en/concept-1/the-piqr-1/

Wynright: Robotic Truck Unloader information. Available online: http://robotics.wynright.com/pdf/RTU.pdf,2013

RobLog industrial demonstrator http://www.roblog.eu/the-roblog-demonstrators/index.html

RobLog Consortium "MS 4 System Operational Process Description" Contract number: FP7- 270350

European Commission, "Technology Readiness Level of Horizon 2020 Work Programme 2014-2015", 2014

A. Urru et al. "Autonomous unloading of heavy deformable goods: Market opportunities" Service Operations And Logistics, And Informatics (SOLI), 2015 IEEE International Conference pp. 54–59 DOI 10.1109/SOLI.2015.7367411 15-17 nov 2015

M.Y. Khan, "Theory & Problems in Financial Management". Boston, 1993: McGraw Hill Higher Education. ISBN 978-0-07-463683-1

Canadian Association of Logistics Management, 12 Feb, 1998

H.B. Maynard, G. Stegemerten, S. Lowry, "Methods Time Measurement", McGraw-Hill, New York, 1948

G. Hoffman "Evaluating fluency in human-robot collaboration - International conference on human-robot interaction (HRI), workshop on human robot collaboration vol. 381 pp 1-8 2013

S. B. Bhandari, "Discounted Payback Period - A Viable Complement to Net Present Value for Projects with Conventional Cash Flows," The Journal of Cost Analysis, Summer 1989, pp. 43–53

M. Bonini et al. "Evaluating Investments in Emerging Automation Solutions for Logistics" in "Innovations and Strategies for Logistics and Supply Chains" pp. 359–388 ISBN: 978-3-7375-4057-5, 2015.

Potential of current digital technologies in the intralogistics sector for SMEs

Raphael Vogt[a], Nur Budi Mulyono[b], Vera Hummel[a], Jan Schuhmacher[a]

[a] ESB Business School, Reutlingen University, D-72762 Reutlingen
[b] School of Business and Management, Institut Teknologi Bandung, 40116 Bandung, Indonesia

---

Abstract

The level of automation in intralogistics has steadily increased over recent years. For small and medium-sized enterprises (SMEs), however, the associated digital change is a major challenge. Since most SMEs are facing increasing sales volumes (e.g. due to e-commerce and a good overall economy) in combination with decreasing lot sizes due to the market demand for individualized products, SMEs have to find innovative solutions to cope with these challenges in production as well as in logistics. Innovative technologies, like 3D printing technologies for the production of small lot sizes and future-oriented intralogistics technologies can serve as enablers in logistics to realize flexible logistic processes for increasing market requirements. Considering that, this paper examines innovative and future-oriented technologies for intralogistics such as smart containers, driverless forklift systems, data glasses, smart shelves and smart pallets regarding their potential for SMEs. This explorative research paper shows that digital technologies are already suitable and available for SMEs. However, challenges are still seen in areas like the identification of digitalization potential and the financing of these new projects. The primary reason especially for SMES for this is that they have to make investments based on an economically feasible payback period and less based on prestigious reasons like digitalization flagship projects done by large corporations. In addition, the identification of feasible starting points for digitalization within intralogistic systems embedded in specific factory processes is a major challenge not only for SMEs.

---

1 Introduction

SMEs are a major pillar of the German economy. According to the Federal Statistical Office 99,3 percent of the companies are assigned to the category of SMEs in Germany (Statistisches Bundesamt, 2014). Often SMEs are not able to conduct extensive future-oriented research because of a certain lack of financial and human resources. In addition, the lack of specific knowhow and missing benefit transparency in combination with issues to identify suitable solutions are key obstacles for digitization of SMEs (BMWi, 2015). This is one major reason for the importance that new technologies have to become accessible, evaluable and affordable for SMEs. On the other hand, the ongoing globalization of markets and an increasing demand for personalized products and (logistic) solutions requires radical changes within supply chains as well as within individual factories. The place of production will undergo constant transformation and the logistical networks with their sources and sinks must be able to follow the adaptability of these production systems. New technologies, the increasing availability of information and communications technology (ICT), a pronounced networking within factories and across the entire supply chain by means of Internet technologies as well as direct and intuitive interaction and collaboration between man and machine offer the potential to proactively counteract these developments and make SMEs ready for the future (Ten Hompel and Henke, 2014).

The following data dealing with digitalization in logistics has been provided by Bitkom (Bitkom e.V., 2017), Germany's digital association, and has been created based on a survey of 508 German companies with more than 50 employees. The vast majority of companies see a variety of advantages through digital technologies in logistics. Around 88 percent perceive the digitalization as an opportunity and 11 percent as a risk. Figure 1 shows the current deployment and planned (future) application of digital technologies in the logistics area of the surveyed companies. The most popular technology already in use is the smart container. Already 20 percent of the interviewed companies are using this technology. Another 15 percent are planning to do so within the next years. Today's use of driverless forklift systems with 19 percent is also quite significant and every fourth company (26 percent) is planning to use this technology in the future. Data glasses for employees (for example to achieve higher efficiency in order picking) are only implemented by 8 percent of the companies and 14 percent are already planning the application. Smart shelves are used by 6 percent of the surveyed companies and another 11 percent are thinking about a future implementation. The deployment of smart pallets is indicated with 4 percent and 10 percent want to introduce them in the future.

The numbers of implemented technologies clearly show that the digital change has already begun in the intralogistics sector. According to a survey conducted by the Federation of German Industries the level of digitalization of SMEs is on average on a level of three on a scale of one to five (Bundesverband der Deutschen Industrie e.V, 2015). For this reason, SMEs still have to continue to face the challenges of digitalization to be prepared for the increasingly fierce international competition. A major key success factor is to foster the exchange between established companies and global players with medium-sized companies and start-ups with innovative ideas for the development of future-orient logistics solutions.

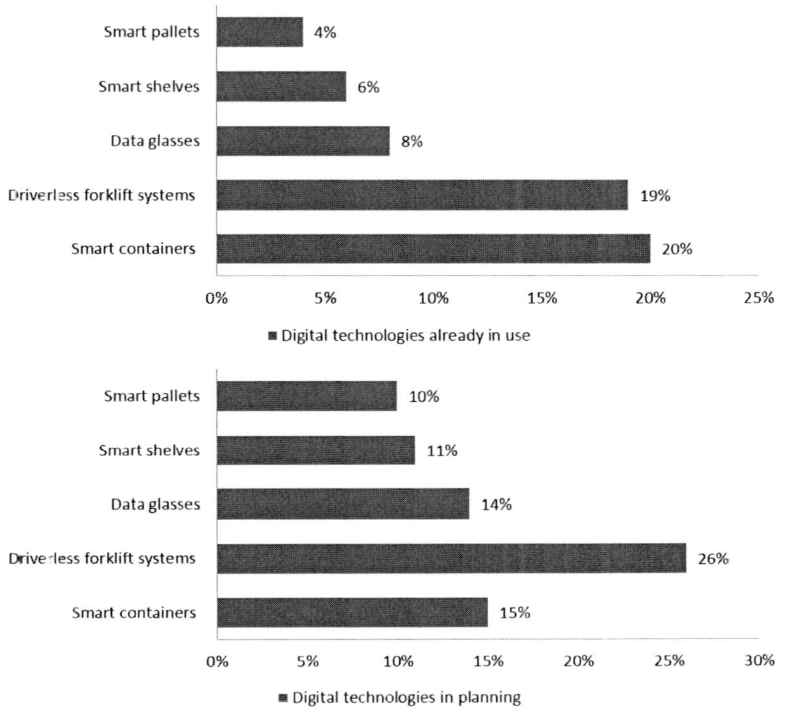

Figure 1: Digital technologies in logistics (cf. Bitkom e.V., 2017). This diagram shows the current deployment and planned future application of digital technologies in the field of logistics of 508 interviewed companies. Smart containers, driverless forklift systems, data glasses, smart shelves and smart pallets have been investigated.

Motivation

A study conducted by Bitkom e.V. shows that the greatest challenges for the companies with logistics processes are the high fuel and energy prices (85 percent) followed by the ongoing digitalization of logistics (74 percent). The hurdles of digital changes are seen not only in terms of costs. Furthermore, the lack of specialist on the market as well as missing standards for digital data transfer have to be addressed and solved (Bitkom e.V., 2017).

Figure 2 shows the digitalization level of German SMEs comparing different regions. Companies with an already high level of digitalization are mainly located in the western (36,8 percent) and northern (30 percent) part of Germany. In the eastern part of Germany, only 23 percent have a high and 35,3 percent a moderate level of digitalization. In Bavaria and Baden-Wuerttemberg the share of SMEs with a high level

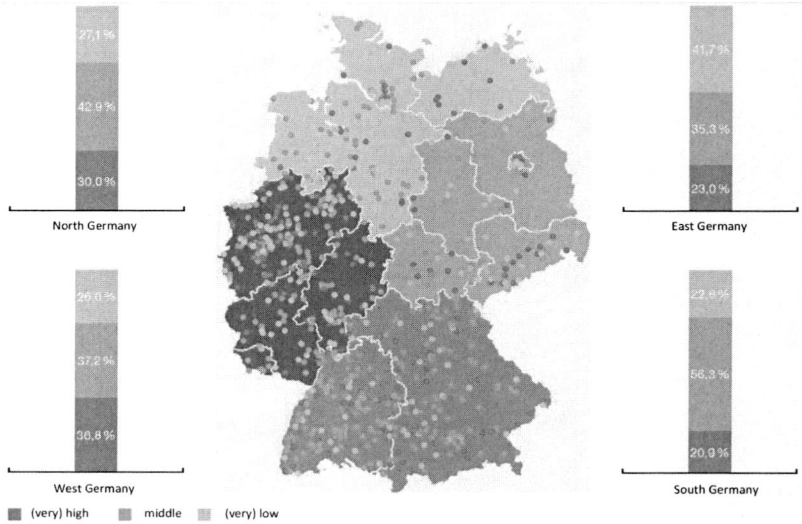

Figure 2: Digitalization level of German SMEs (cf. Bundesverband der Deutschen Industrie e.V, 2015). This figure shows the digitalization level of German SMEs comparing the regions North, South, East and West Germany.

of digitalization is even lower (20,9 percent). The high proportion of companies with an average digitization level restricts the performance of the companies in the South. It is striking that industrial companies want to optimize their sales opportunities with new digitization projects (Bundesverband der Deutschen Industrie e.V, 2015).

Therefore, a crucial precondition for a successful digitization of SMEs is to examine the potential of digital technologies to identify those technologies, which generate the greatest benefit. As the digitalization level of German SMEs (figure 2) also illustrates there is still a lot to be done to develop future-oriented digitalization strategies. In this way, concrete suggestions for the introduction of advanced intralogistics technologies have to be designed.

2 Approach

In order to estimate the potential of selected digital technologies for SMEs, the state-of-the-art technologies "Smart container", "Driverless forklift systems", "Smart glasses", "Smart shelves" and "Smart pallets" identified as key technologies will be described in more detail and assessed based on three different factors: accessibility, suitability and affordability.

Accessibility: Many technologies are so far only accessible to established companies and global players. This may be due to patent law, ongoing research projects or due to the current technical development status. If the technology is not yet fully developed, the time of market maturity is taken into account.

Suitability: It is considered which particular technical requirements must be provided for the implementation of the technology in the intralogistics and warehouse logistics process. How must the work environment look like to ensure a successful exchange of the new technology with already existing systems? Furthermore, a suitable interface design has to be found.

Affordability: For this factor, the initial investments for the company are assessed. In addition, the risk of subsequent costs is tried to be answered. Finally, the case of a cost increase is considered by a transformation of the company's internal infrastructure.

3 State of the Art

Today's material flow systems are designed based on complex, centralized material flow controller architectures programmed based on predefined processes which would lead to a huge increase of complexity to deal with small lot sizes down to lot size 1 orders and a constant programming effort in changeable production systems (Günthner et al., 2014; Wilke, 2006). To achieve future flexible logistic processes a decentralization and autonomy of the logistic decision-making and control processes are required. Based on new information and communication technologies which are applied in intelligent logistical infrastructure and which are able to take over planning and control processes these duties can be transferred to the level of the physical material flow (Scholz-Reiter et al., 2005). The development and application of decentralized control methods for material flow systems combined with the approach of the Internet of Things bear a huge potential to solve the arising challenges within future production systems. A cornerstone for this is that every automated material flow system can be structured with the basic units of the Internet of Things which are cooperating functional units (also called entities) of conveyor modules, transport units and (software) services (Günthner et al., 2014). So transport units respectively bins, intermodal container etc. are becoming autonomous entities which guide themselves to their target destination, whereas intelligent conveyor systems, like driverless forklift systems, are executing the actual transport of the transport units by communicating with other entities and by using the (software) services of the cyber-physical production system or of the supply chain network.

In the following sub-sections, the technical bases for the evaluation are elaborated. The basic description is not intended to be specific to one product. Thus, if several products from different suppliers are already available on the market, selected technical features of the products will be included in the evaluation.

3.1 Smart container

Besides the conveyor systems, also the logistical transport units have to evolve to become part of the emerging Internet of Things. This can be done by transferring more intelligence to control logistics processes directly to the transport unit by adding computing capacities, identification capabilities, decision-making capabilities/logics, sensors and capabilities for wireless communication to these logistical objects. One of the best-known intelligent transport units is the smart container. As already mentioned in the introduction, this technology is already used quite often in companies. Different versions of the smart container are already available on the market. However, the term "smart container" is used for intelligent load carriers on the intralogistics level (like bins and small load carriers) as well as for ISO intermodal containers for transport on ocean-going vessels on the intercontinental level (Lang et al., 2011). The main application area of smart containers for intralogistics is primarily located in the C-part management. The container can be equipped with a certain level of intelligence depending on the operational scenario. Straube et al. (2016, pp. 61 f.) state that smart containers make complex logistical processes more manageable and thus decentralization of planning and control decisions can be achieved.

The containers can be integrated into an existing infrastructure via radio frequencies such as 868 MHz or 2.4 GHz and the 6LoWPAN communication protocol. Smart containers usually consist of the following units: graphic display, control buttons, tracking system, transmitting and receiving unit, charging controller, solar cell, energy storage, microprocessor and storage unit (Roidl et al., 2014). This enables an energy self-sufficient communication between the containers. Furthermore, according to Roidl et al. (2014, p. 2) it is possible to intervene the information exchange via the graphic display and the control buttons. Intelligent containers allow a decentralization of data storage following the data-on-the-chip principle by storing all relevant information of the transport unit like a unique identification number, the source and sink information of the bin, content information and the dimensions or geometric shape of the bin directly on the microcontroller of the transport unit. The coupling of information and material flow also reduces the required communication of logistics related information and data redundancy within the system. In addition, tasks as routing, transport and avoidance of congestions can be fulfilled decentrally and autonomously by the intelligent bins in cooperation with other logistic entities in the Internet of Things without a central control system (Günthner et al., 2010).

3.2 Driverless forklift system

Driverless forklift systems are used for the automatic transport of pallets, containers and other goods in the warehouse. The the implementation of driverless forklift systems is particularly suitable, if the following conditions are present (Schulte, 2016, pp. 215-219):

- High picking volume
- Along a fixed route

Variants	Description
Mechanical constrain	Grooves and rails
Conduction-oriented guidance	Optical ground lines, passive steel strips and induction guide wires
Wireless forcible control	Laser navigation, coordinate systems, ambient scanning and GPS technology

Table 1: Variants for automatic vehicle guidance (Schulte, 2016, pp. 215-219)

- Foreseeable goods movements

Depending on the application scenario, either a purely autonomous or a combination of manual (with driver) and autonomous forklift system can be used. Starting from a dual-mode forklift, the following system components are installed: vehicle control system, drive control, motor, automation module, safety module, navigation module and data transmission module. (Crown Gabelstapler GmbH & Co. KG, 2013)

A major challenge for developers is currently to secure indoor navigation, especially within rapidly changing environments and areas that are also used by employees. Table 1 shows the three most common approaches for automated guided vehicles. The optical line navigation is currently the most widely used variant. In future navigation with natural landmarks as well as *Simultaneous Localization and Mapping* (*SLAM*) approaches will play an increasingly important role to cope with changing factory environments.

3.3 Smart glasses

In the future, many companies will rely on smart glasses to assist manual picking process. The so-called „pick-by-vision" should lead to an increase in ergonomics and efficiency (Schulte, 2016, pp. 279 f.). In addition, real-time data can be included in the picking process.

Major advantages if the picker wears smart glasses are that he has both hands free during the picking process and he is also provided with the necessary information for picking (like storage location, item description, quantity...) via an integrated display. Essential prerequisites for the use of smart glasses in warehouse locations are a comprehensive WLAN network as well as sufficient light for the scanning processes (Schulte, 2016, pp. 279 f.). Also real-time data transfer is a crucial precondition for the use of smart glasses in combination with a warehouse management system (or enterprise resource planning system) since all goods movements and stocks always have to be up-to-date in the system to achieve the full potential of these solutions. Head-mounted displays have proven to be particularly handy, because the virtual information appears in the field of view but the worker still can see the real environment in the background (Imhof and Thalmann, 2016, pp. 3 f.).

3.4 Smart shelves

A specific version of an intelligent shelf uses lockable cabinets. This solves the problem of access by employees via the use of a person-specific RFID card. Thus, each time the cabinet is opened the time of access and the RFID card number are registered. The inventory of each individual product or container is recorded by the use of electronic scales. The scale is integrated in the shelves. Any weight change is registered and the recorded data is sent to a controller. The information on the controller is filtered and forwarded to the server via LAN, WLAN or GSM (Global System for Mobile Communications) connection. In this way, each removed part is documented and can be assigned automatically to a cost center. The system defines product-specific minimum stocks. If there are fewer parts than the minimum stock in the shelf, a reorder is sent automatically (DIGI SENS AG, 2016). This type of smart shelf can be a cost-effective alternative to smart containers, since conventional bins can still be used to monitor the container filling level. Furthermore, a popular radio protocol for intelligent shelves is the ZigBee. A great advantage of the ZigBee protocol is that each network user can also be used as a router and enabling a very stable and reliable communication (Mettler-Toledo AG, 2012).

3.5 Smart pallets

Smart pallets are equipped with transponders (such as RFID tags), GSM/WLAN/GPS-unit, processing unit, battery and connections for additional sensors. If the respective hard-and software is used, the location, movement or other parameters like temperature, weight and shocks of the pallets can be monitored (Bullinger and ten Hompel, 2007, pp. 83 ff).

According to the Whitepaper from Zebra technology (2015, p. 3-6), the advantages of RFID pallets are:

- a trouble-free load identification without a visual connection
- an easy detection of inventory
- the reliable positioning of pallets with load
- a faster handling of inventory
- the punctual collection of goods movements (tracking)

The RFID technology also enables bulk recognition in metal environment. The following different types of load unit detection based on RFID are currently used in industry:

4 Study of technologies

The following sub-sections examine the potential of the five described technologies for use in small and medium-sized enterprises based on their accessibility, suitability

Variants	Description
Handheld	The operator scans all transponders attached to the frames and pallets with a handheld reader.
Gate	The unit pack IDs of the pallet transponders are read with a gate reader attached to the gate.
Forklift trucks	Forklift trucks readers are mainly used when (un-)loading units into trucks.

Table 2: Types of detection (RFID) (Bullinger and ten Hompel, 2007, pp. 128 ff)

and affordability.

4.1 Accessibility

The accessibility of the technologies is to be assessed with the help of the measurement system Technology Readiness Level. It is a method developed by NASA (National Aeronautics and Space Administration) to assess the level of maturity of a new technology. There are nine different technology readiness levels. TRL 1 (Basic principles observed and reported) represents the lowest and TRL 9 (Actual system "flight proven" through successful mission operations) the highest level (National Aeronautics and Space Administrations, 2015).

The TRL measurement system is a well-established measurement system in EU research and is therefore a common method for assessing new technologies (European Commission, 2014). Figure 3 shows the 9 TLR levels with the corresponding explanations. The assignment of the technologies to the TRL level has been done based on an internet research. An important guideline is the estimated time for the market launch of a technology. The goal of the internet research was to find companies or research instituts, which have already successfully developed a product. The products are then examined for their development status (time to market launch) and assigned to a TRL level.

The inBin developed by the Frauenhofer Institute for Material Flow and Logistics (IML) and the iBin of Würth Industrie Service are used to assess the smart container technology. It should be taken into account that by the accession of the Würth Industrie Service to the Enterprise Lab Center (opened by the Frauenhofer IML) the linking of the two products iBin and inBin is possible (Frauenhofer IML, 2016, p. 28). The smart container technology is classified as TLR 9 since there are already developed products available on the market, which can be bought from various suppliers.

For the investigation of driverless forklift systems, a reach truck developed by Jungheinrich AG and the Institute for Integrated Production Hannover (IPH) is used as representative example. By the use of innovative hardware and software, the reach truck can interpret the human language and gestures (IPH gGmbH, 2016). This level of intelligence is new in the product range of driverless forklift systems. The installed

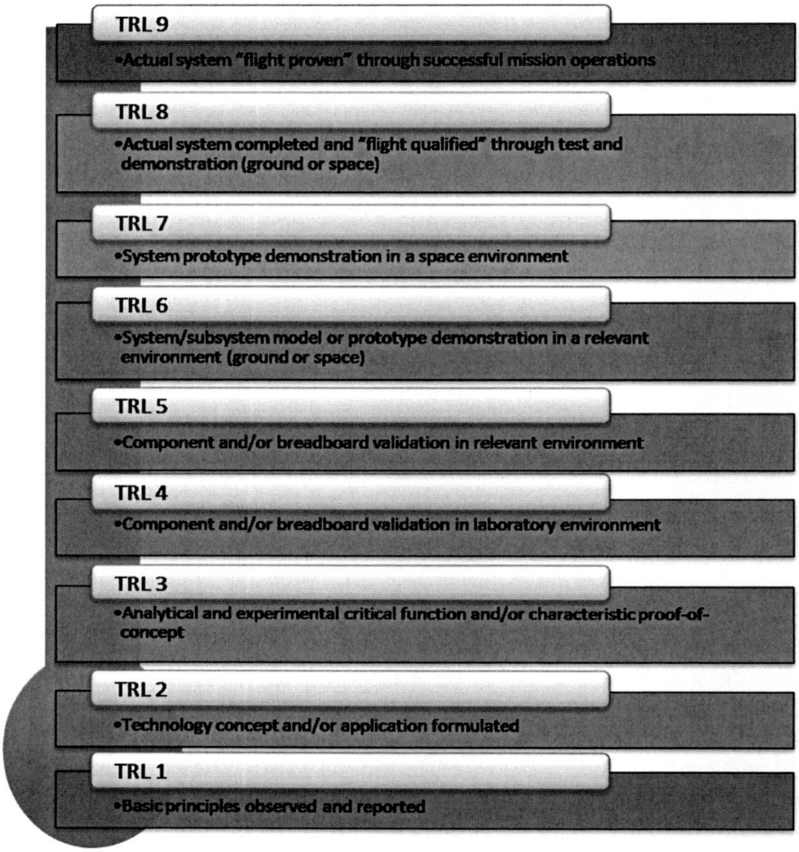

Figure 3: Technology Readiness Level (National Aeronautics and Space Administrations, 2015). The figure shows the different levels of the measurement system Technology Readiness Level. The TRL describes the stage of development of a technology, procedure or service.

3D cameras of the reach truck capture the environment and ensure a safe navigation based on distinctive points in the factory environment (IPH gGmbH, 2016). So far, this smart reach truck cannot be purchased yet. However, there are suppliers offering driverless forklift systems with a lower level of intelligence (euronews, 2015). However, if one considers the described reach truck, the product is currently classified as TRL 7 (System prototype demonstration in a space environment).

According to Picavi GmbH, they are the first company to offer a marketable product for picking with smart glasses (picavi GmbH, 2017). In order to ensure the continuous operation of the data glasses, the Picavi Power Control (battery) provides sufficient energy for at least one shift (picavi GmbH, 2017). In addition, Picavi provides the data glasses with the necessary software. Since the middle of the year 2015, the technology for warehousing processes (incoming goods, picking, outgoing goods and inventory) has been in permanent use in several companies (picavi GmbH, 2017). Based on the already successful market launch the smart glasses are assessed with a TRL 9. It must be noted that many smart glasses solutions for warehousing process are still in TRL 6 to TLR 8.

A smart shelf offered by Mettler-Toledo AG uses weighing sensors integrated into the shelves (Mettler-Toledo AG, 2012). This modular design facilitates different application scenarios, like the integration in shelves, on mobile racks, in jobsite containers and in industrial vending machines (Mettler-Toledo AG, 2012). By using existing digital networks, the weight of the containers or the material on the shelf can be checked remotely (Hrebenciuc et al., 2011). The measured weight is then used to determine the actual stock level in real-time and if a certain stock level is reached a replenishment order can be placed automatically. Also it does not matter what material is on the shelf, so the system can be used for different industries e.g. for digitalization of stock keeping of critical parts. Various smart shelf can already be purchased which leads to a TRL 9 classification.

In 2013, the pallet rental company Chep and the Fraunhofer Institute for Material Flow and Logistics tested the first intelligent goods carriers in the SmaRTI project (Smart Reusable Transport Items) (Hector and Knorre, 2015). In this cloud-based tracking & tracing system, data can be collected at various points in the supply chain and events can be generated in real time (Anderseck and Hille, 2013). The project partners Chep, Mars, Rewe and Fraunhofer IML have already carried out the first practical test successfully (Hector and Knorre, 2015). The industrial application of this technology leads to a TRL 9.

Depending on the suppliers, the market and on the level of intelligence various different products can be found. For all five technologies, there is at least one supplier which already sells a marketable product with a lower degree of intelligence. This means that these products in general are classified as TRL 9. For some technologies mentioned in this paper, a little caution is required since the product has only recently been introduced to the market.

4.2 Suitability

In order to assess the suitability for SMEs, the technical requirements for a product implementation are considered. Common standards, interfaces and reference architectures like the "Reference Architecture Model Industrie 4.0 (RAMI4.0)" (DIN SPEC 91345, 2016) can make a major contribution to simplify the implementation of these new systems. In the following, a brief insight into the suitability of the selected technologies for SMEs should be given by examining the five technologies for the following three basic technical requirements:

Compatibility: Standardized interfaces are required for the integration into existing systems.

Functional reliability: The technology must have been thoroughly tested for its viability in a productive environment, to ensure that it is free of bugs and security gaps.

Modifiability: The introduced technology must be adaptable to changing requirements at reasonable costs.

The technologies are assessed using the following scale:

+	Available
(+)	Available with certain restrictions
–	Not available
∅	No prediction possible

Technical requirements	Smart containers	Driverless forklift systems	Data glasses	Smart shelves	Smart pallets
Compatibility	+	+	(+)	(+)	+
Functional reliability	+	–	+	+	+
Modifiability	+	+	∅	+	+

Table 3: Assessment of the technologies

In most cases, the suitability of the investigated technologies for SMEs is not an obstacle. In the case of the autonomous forklift system, functional reliability is limited by the not yet completed market launch of highly autonomous systems. However, the current development status (several tests have already been carried out) suggests that it will be launched in the near future.

4.3 Affordability

The following assessment of the total cost of a product implementation are based on a student survey of ESB Business School students from the Digital Industrial Management course. In the lecture Research Methods and Planning, conducted by Professor Nur Budi Mulyono, the technologies were classified as follows in figure 4. The diagram shows the estimated total cost on the y-axis and the estimated adaptation effort of the infrastructure on the x-axis. The estimated adaptation of the infrastructure refers to the modifications, which are necessary in order of a technology rollout in the warehouse.

The smart container can easily be integrated into an existing infrastructure without much additional effort and costs. The startup time is therefore also manageable. In addition, the existing shelf systems and work processes in the warehouse can be maintained. In future, smart containers will be available for less than 20 euros each on

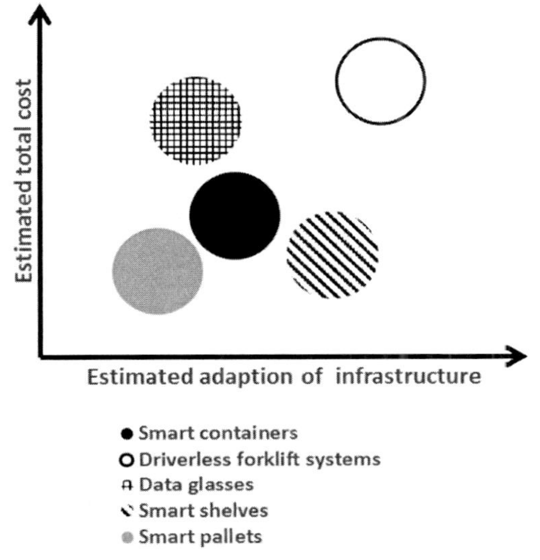

Figure 4: Affordability of digital technologies for SMEs. This diagram shows the estimated total cost and the estimated adaptation effort of the infrastructure. Smart containers, driverless forklift systems, data glasses, smart shelves and smart pallets are investigated.

the market depending on the required functionalities and level of autonomy (Gabel, 2013). The total cost of a container system depends, of course, on the number of needed smart containers, the used technology and especially on the required functionalities and capabilities as well as on the specific factory environment or supply chain in which it is to be implemented.

Considering the acquisition unit costs the forklift truck which is capable to navigate in free space is the largest investment compared to the other four technologies. In contrast to previous driverless transport systems, however, no predetermined routes have to be defined (IPH gGmbH, 2016). So the routes in the factories does not have to be marked with magnetic or optical tracks on the ground which are required for navigation for most conventional automated guided vehicles, which saves costs for installation, maintenance and adaption of the tracks. These circumstances can make this new technology economically profitable for SMEs. However, the non-optimal spatial division of the warehouses among SMEs could be an obstacle for existing warehouses. Other challenges for the implementation of driverless forklift systems are usually to be found in internal transport routes which are too long, narrow widths of the transport passages and the use of inappropriate equipment (Hohberger and Damlachi, 2014, pp. 166).

The acquisition cost associated with the introduction of data glasses into the picking process are up to fifty percent lower than the introduction of conventional systems

(TRENDS Business Development, 2016). For example, a pick-by-light system requires a display at every storage location. In a pick-by-vision system, only every picker needs a data glass and the storage locations only have to be marked with markings for storage location identification like barcodes.

Most of the smart shelf solutions fit into standard racking systems. For example, up to eight scales can be installed on a shelf of the Mettler-Toledo GmbH. The profitability forecast predicted rapid amortization and staff savings due to rationalization effects. It is assumed that the system will return the acquisition value thanks to time savings, simplified processes, less costs due to errors and reduced printing costs in less than two years. (Mettler-Toledo GmbH, n.y., p. 2.)

The implementation of smart pallets is to be simplified by means of standardized development architectures for AutoID technologies and IT services. The cost of a smart pallet is usually not significantly higher compared to a standard pallet. By reducing the number of process errors the use of smart pallets can save up to 8 billion euro in the German trade. In addition, pallet losses can be reduced by up to 50 percent (Spehr, 2013, p. 10).

5 Conclusion

The area of logistics plays an increasingly important role in the annual planning of SMEs. The change from a purely transport oriented to a performance determining instance and competitive factor illustrates this. It turns out that the manufacturing industry adapts more and more to the technical requirements of logistics processes (Auffermann, n.y., p. 6.). With regard to the technological development in intralogistics in context of Industrie 4.0, intelligent systems to support employees will become increasingly important. If additional information on current market-available digital technologies is needed, SMEs can make use of the investigations in this article. Compared to the high potential due to improvements in the warehouse logistics process, the degree of conversion is still rather small. Most of the required technologies have already been extensively developed and tested. This is also confirmed by the study carried out. The support of SMEs with the introduction and evaluation of intelligent systems for their specific needs could be a possible solution to increase the use innovative technologies in logistics leading to more efficient processes and cost savings. SMEs' concerns about new digital solutions need to be addressed through concrete and successfully implemented projects. In this way, the added value of a connected factory environment can be proven in practice. An important factor will also be the use of standardized interfaces and architectures of the applied technologies, not only for single companies but entire value chains. This leads to the conclusion that the five considered technologies are available and suitable for SMEs in general. However, problems are still seen in the financial evaluation of these new technologies. This is of high importance especially for SMEs since they only can make investments with a predictable amortization period in order to remain competitive. E.g. if a subarea is digitized in the warehouse logistics process, many SMEs are missing the corresponding financial reserves to implement follow-up projects if the cost savings are not sufficient to cover the investment costs within a short period of time. Also the entire logistic infrastructure usually can only be adapted medium to long term, especially if they are closely linked to specific supply chains or industries, like the automotive industry, with existing standards. However, the potential of digitization is recognized by most SMEs as outlined in this paper. According to the *Mittelstand 4.0 Kompetenzzentrum Dortmund* the framework conditions for SMEs are ideal for industrial projects. High flexibility and flat hierarchies often characterize the company structure of SMEs, which is the best prerequisite for digital transformation and a huge advantage in comparison to huge multinational corporations (Hermanski and Parlings, 2016, p. 2).

Acknowledgement: This scientific paper was created as part of the international research master *Digital Industrial Management and Engineering* at the ESB Business School.

Literature

Anderseck, B. and Hille, A. (2013): smaRTI: Deploying the Internet of Things in Retail Supply Chains, Dortmund.

Auffermann, C., Kamagaev, A., Nettsträter, A., ten Hompel, M., Verbeek, K. and Wolf, O. (n.y.): Thesenpapier: Cyberphysical Systems in der Logistik, EffizienzCluster Management GmbH, Mühlheim an der Ruhr.

Bitkom e.V., editor (2017): Digitalisierung der Logistik, Berlin.

Bullinger, H. and ten Hompel, M. (2007): Internet der Dinge, Berlin, Springer-Verlag.

Bundesministerium für Wirtschaft und Energie (BMWi), editor (2015): Industrie 4.0, Volks- und betriebswirtschaftliche Faktoren für den Standort Deutschland, Eine Studie im Rahmen der Begleitforschung zum Technologieprogramm AUTONOMIK für Industrie 4.0, Berlin.

Bundesverband der Deutschen Industrie e.V. and PricewaterhouseCoopers AG, editor (2015): BDI/PwC-Mittelstandspanel – Digitalisierung im Mittelstand, Berlin.

Crown Gabelstapler GmbH & Co. KG, editor (2013): Automatisierung von Gabelstaplern: Sind wir schon so weit?, München.

DIGI SENS AG, editor (2016): Das intelligente Schrank-System - i.cupboard, Murten.

European Commission, editor (2014): Horizon 2020 – Work Programme 2014 – 2015.

Euronews, editor (2015): Fahrerlose Gabelstapler, Washington. Retrieved from http://de.euronews.com/2015/10/12/fahrerlose-gabelstapler (accessed on September 09, 2017)

Frauenhofer IML, editor (2016): LOGISTIKentdecken Magazin des Frauenhofer-Instituts für Materialfluss und Logistik IML Dortmund 16, Dortmund.

Gabel, O. (2013): Hohenloher Tagesblatt: Den Bestand fest im Blick, Bad Mergentheim. Retrieved from http://www.swp.de/crailsheim/lokales/region/den-bestand-fest-im-blick-6699362.html (accessed on September 09, 2017)

Günthner, WA., Chisu, R., Kuzmany, F. (2010): Die Vision vom Internet der Dinge. In: Günthner W, ten Hompel M, editors. Internet der Dinge in der Intralogistik. Berlin: Springer; 2010.

Günthner W, Klenk E, Tenerowicz-Wirth P. (2014): Adaptive Logistiksysteme als Wegbereiter der Industrie 4.0. In: Bauernhansl T. et al. (editors), Industrie 4.0 in Produktion, Automatisierung und Logistik, Springer Fachmedien Wiesbaden 2014

Hector, B. and Knorre, J. (2015): Paletten-Handbuch Tausch – Rechtspraxis – Trends, Hamburg, DVV Media Group GmbH.

Hermanski, P. and Parlings, M. (2016): Das Kompetenzzentrum für den Mittelstand: Einfach näher dran, Forschungs- und Innovationsreport Industrie 4.0, Dortmund.

Hohberger, S. and Damlachi, H. (2014): Praxishandbuch Sanierung im Mittelstand, Wiesbaden, Springer Gabler Verlag.

Hrebenciuc, F., Stroia, N., Moga, D. and Barabas, Z. (2011): A Low Cost Approach to Large Smart Shelf Setups, Romania.

Imhof, S. and Thalmann, A. (2016): Head-Mounted Displays – Arbeitshilfen der Zukunft Bedingungen für den sicheren und ergonomischen Einsatz monokularer Systeme, Dortmund.

Institut für Integrierte Produktion Hannover gGmbH IPH, editor (2016): Presseinformation: Hannover Messe 2016: Intelligenter Stapler versteht Sprache und Gesten, Hannover.

Lang, W., Jedermann, R., Mrugala, D., Jabbari, A., Krieg-Brückner, B. and Schill K. (2011): The "Intelligent Container"—A Cognitive Sensor Network for Transport Management. IEEE Sensors Journal, Volume 11 (2011), NO. 3, p. 688-698.

Mettler-Toledo AG, editor (2012): Weighing Components for Remote Inventory Solutions, Switzerland, Greifensee.

Mettler-Toledo GmbH, editor (n.y.): New: Produktion Industrielles Wägen und Messen Teile rationell brewirtschaftenk, Giessen.

National Aeronautics and Space Administrations, editor (2015): Technology Readiness Level, Washington. Retrieved from https://www.nasa.gov/directorates/heo/scan/engineering/technology/txt_accordion1.html (accessed on September 09, 2017) picavi GmbH, editor (2017): Kommissionieren mit Datenbrille! Logistik 4.0 im Lager, Herzogenrath. Retrieved from http://picavi.com (accessed on September 09, 2017)

Roidl, M., Emmerich, J., Masoudinejad, M., Riesner, A. and ten Hompel, M. (2014): Entwicklung eines Versuchsfelds für große Systeme intelligenter Behälter, Dortmund.

Scholz-Reiter B, Freitag M, Rekersbrink H, Wenning BL, Gorldt C, Echelmeyer W (2005): Auf dem Weg zur Selbststeuerung in der Logistik-Grundlagenforschung und Praxisprojekte. In: G. Wäscher et al., editors. Intelligente Logistikprozesse. Konzepte, Lösungen, Erfahrungen. Companion volue "11. Magdeburger Logistiktagung". Magdeburg: LOGiSCH 2005.

Schulte, C. (2016): Logistik – Wege zur Optimierung der Supply Chain, München: Vahlen Verlag.

Spehr, H., Pannicke, M., Sandlöhken, J. and Anderseck, B. (2013): ECR-Award 2013: smaRTI in der Konsumgüterindustrie (Die Revolution des Material-und Datenflusses).

Statistisches Bundesamt – Destatis, editor (2014): Anteile kleiner und mittlerer Unternehmen an ausgewählten Merkmalen, Wiesbaden.

Straube, F., Baumgarten, H. and Klinkner, R. (2016): Technologieorientiertes Service Engineering in der Kontraktlogistik – Konzeption eines Gestaltungsmodells für die systematische Entwicklung technologiebasierter Logistikdienstleistungen, Berlin, Universitätsverlag der TU Berlin.

Ten Hompel, M., Henke, M. (2014): Logistik 4.0, in: T. Bauernhansl et al. (editors), Industrie 4.0 in Produktion, Automatisierung und Logistik, Springer Fachmedien Wiesbaden 2014

TRENDS Business Development, editor (2016): Intralogistik: Pick-by-Vision – Smart Glasses in der Kommissionierug, München. Retrieved from http://intralogistik.tips/pick-by-vision-smart-glasses-in-der-kommissionierung/ (accessed on September 09, 2017)

Wilke M. (2006): Wandelbare automatisierte Materialflusssysteme für dynamische Produktionsstrukturen. Dissertation. Munich: Herbert Utz Verlag GmbH 2006.

Zebra Technologies, editor (2015): Managed smart pallet service: Improving supply-chain efficiency while driving down costs, Switzerland.

Beiträge in Deutscher Sprache

Planungsvorgehen für die Digitale Fabrikplanung im Kontext der Digitalen Fabrik

Harald Augustin[a]

[a] *ESB Business School, Reutlingen University, D-72762 Reutlingen*

Abstract

Die Digitalisierung verändert die Arbeitsweisen in Unternehmen und damit auch bei der Planung von Fabriken. Am Markt verfügbar sind vielfältige Soft- und Hardwaresysteme für die Digitale Fabrik, die die Fabrikplanung unterstützen und teilweise revolutionieren können. Um dieses Potential zu nutzen, muss aber bekannt sein, was diese Systeme können und wie sie sinnvoll genutzt werden können. Bisherige Fabrikplanungssystematiken haben hierzu keine Unterstützung angeboten, da sie nur das generelle Vorgehen der Fabrikplanung beschreiben ohne den Zusammenhang zu den sich rasch entwickelnden Soft- und Hardwaresystemen herzustellen. Im Folgenden wird die Entwicklung und Umsetzung eines Planungsvorgehens für die Digitale Fabrikplanung dargestellt, das die Potentiale der Soft- und Hardwaresysteme für die Digitale Fabrik aufgreift und in ein passendes Planungsvorgehensmodell integriert.

---

1 Einordnung und Ziele der Digitalen Fabrikplanung

1.1 Die Digitale Fabrikplanung als Teil der Digitalen Fabrik

Die Digitale Fabrik wird in ihrem Umfang, ihren Inhalten und Schwerpunkten unterschiedlich definiert, u.a. von folgenden Autoren:

- Nach *Schenk, Wirth und Müller* (*2014*) beinhaltet der Begriff der Digitalen Fabrik den Entwurf von digitalen Modellen zur Planung und Simulation von Produkten, Prozessen, Anlagen sowie Produktions- und Logistiksystemen. Das virtuelle Modell umfasst neben den planerischen und simulativen Gestaltungselementen auch die Analyse sowie Optimierung der Fabriksysteme über den gesamten Lebenszyklus hinweg. (vgl. Schenk, Wirth und Müller, 2014, S. 397, 399)
- *Kühn* (*2006*) spricht im Zusammenhang mit der Digitalen Fabrik von der Bereitstellung digitaler Systeme und Werkzeuge zur Planung und Simulation von

Fabrikstätten. Das Hauptaugenmerk sieht er in der Vernetzung und Integration einer von verschiedenen Unternehmenseinheiten genutzten Datenbasis. (vgl. Kühn, 2006, S. 1)

- *Grundig (2015)* definiert die Digitale Fabrik als ein Netzwerk von digitalen Modellen, Methoden und Werkzeugen zur Planung, Simulation und Visualisierung eines virtuellen Abbildes der realen Fabrik, die durch ein ganzheitliches Datenmanagement miteinander vernetzt sind. (vgl. Grundig, 2015, S. 32)

Eine allgemeingültige Definition bietet die VDI 4499 Blatt 1 (2008): „Die Digitale Fabrik ist der Oberbegriff für ein umfassendes Netzwerk von digitalen Modellen, Methoden und Werkzeugen – u. a. der Simulation und der dreidimensionalen Visualisierung –, die durch ein durchgängiges Datenmanagement integriert werden. Ihr Ziel ist die ganzheitliche Planung, Evaluierung und laufende Verbesserung aller wesentlichen Strukturen, Prozesse und Ressourcen der realen Fabrik in Verbindung mit dem Produkt." (VDI 4499 Blatt 1, 2008, S. 3)

Die Digitale Fabrik hat ihren Fokus in den Lebenszyklusphasen auf die Produktionsplanung und Gestaltung der Fabrik mit Schnittstellen und Übergängen zur Entwicklung und Konstruktion sowie zur Produktion (vgl. Abb. 1; VDI 4499 Blatt 1, 2008, S. 3–4). Die Fabrikplanung beschreibt die Planungsphasen innerhalb der Lebenszyklusphase „Produktionsplanung und Gestaltung der Fabrik" im Detail (vgl. Abb. 1). Der Digitalen Fabrikplanung kommt in der Digitalen Fabrik eine besondere Rolle zu, da sie die Vorgehensmodelle für die Planung im digitalen Kontext bereitstellt. Ebenso definiert sie einzusetzende Methoden und Werkzeuge im Planungsprozess. Nach Abschluss der Planungsphase werden die erzeugten Modelle und Daten im Betrieb der Produktion für die fortlaufende Optimierung eingesetzt.

In VDI 4499 Blatt 1 (2008) und VDI 5200 Blatt 1 (2011) sind zwar die grundlegenden Strukturen der Lebenszyklus- und Planungsphasen beschrieben, allerdings ohne eine strukturelle und inhaltliche Verknüpfung für ein Vorgehensmodell zur Digitalen Fabrikplanung aufzuzeigen. Da die Fabrikplanung unter Einsatz von Methoden und Werkzeugen der Digitalen Fabrik aber einen anderen Charakter hat, ist auch das Vorgehensmodell der Fabrikplanung unter Berücksichtigung der Potentiale der Digitalen Fabrik weiter zu entwickeln. Das heißt, dass die Prozesse und Aktivitäten insbesondere aus der Produktionsplanung und Gestaltung der Fabrik mit ihren spezifischen, auf die Digitale Fabrik ausgerichteten Anwendungen in die Fabrikplanungsphasen, z.B. nach VDI 5200 Blatt 1 (2011), übernommen und integriert werden müssen. Dadurch ändern sich die Planungsphasenmodelle strukturell und werden inhaltlich erweitert. Ein derartiges Planungsphasenmodell der Digitalen Fabrikplanung, das sich an den Zielen der Digitalen Fabrikplanung aus Kapitel 1.2 orientiert, wird in Kapitel 3 vorgestellt.

1.2 Nutzen und Ziele der Digitalen Fabrikplanung

Die Ermittlung eines genauen Kostenvorteils durch den Einsatz von Methoden und Werkzeugen der Digitalen Fabrik lässt sich für ein Unternehmen nur schwer ermitteln,

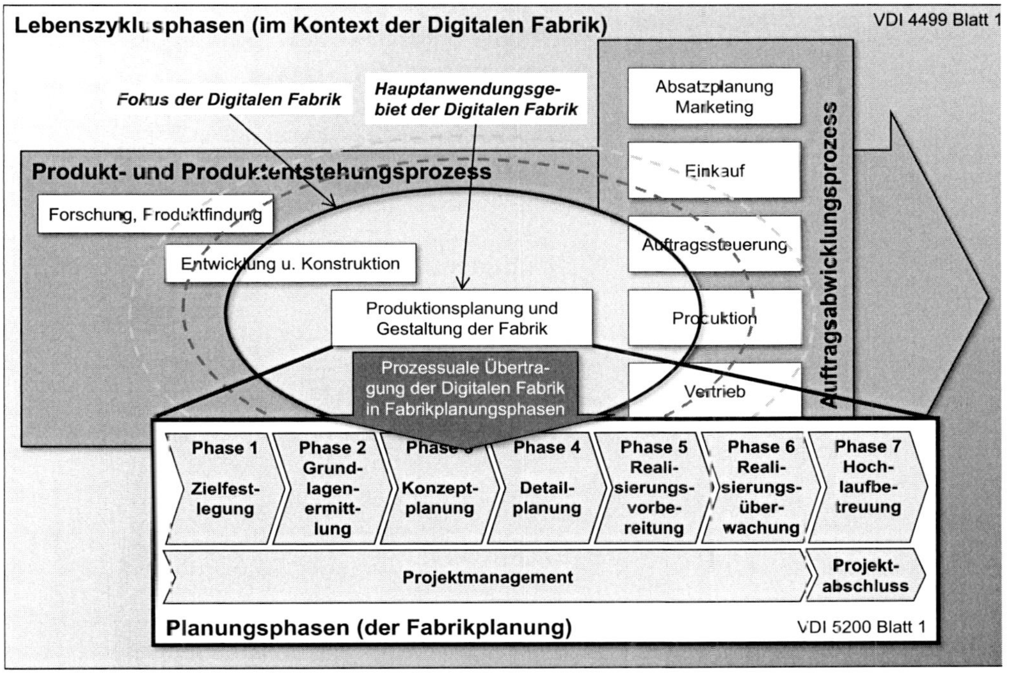

Abbildung 1: Einordnung der Fabrikplanung in den Kontext der Digitalen Fabrik (in Anlehnung an VDI 4499 Blatt 1, 2008, S. 7; VDI 5200 Blatt 1, 2011, S. 9)

da sich mehrere Einflüsse zu einer Gesamtwirkung zusammenfügen. (vgl. Bracht, Geckler und Wenzel, 2011, S. 51) Dies gilt auch für die Digitale Fabrikplanung als zentraler Teil der Digitalen Fabrik. Die Digitale Fabrikplanung kombiniert darüber hinaus den Nutzen einer strukturierten Fabrikplanung und der Digitalen Fabrik. Bei der Planung, Erstellung und Inbetriebnahme der Ressourcen sowie im Fabrikbetrieb werden somit die in Tab. 1 dargestellten Ziele und Nutzen erreicht.

2 Planungsphasenmodelle zur Fabrikplanung im Kontext der Digitalisierung

Verschiedene Autoren stellen in ihren Fabrikplanungssystematiken unterschiedliche Planungsphasenmodelle zur Fabrikplanung in der Literatur dar. Auf einer übergeordneten und damit ersten Prozessebene, die über den Planungsphasen nach VDI 5200 Blatt 1 (2011) liegt, gibt es eine gemeinsame Basisstruktur, in die alle vorliegenden Planungsphasenmodelle zur Fabrikplanung eingeordnet werden. Diese erste Prozessebene umfasst die Prozessschritte Vorbereitung, Strukturplanung, Detailplanung, Ausführungsplanung und Ausführung (vgl. Abb. 2). Ihr folgt die zweite Prozessebene,

Nutzen der Digitalen Fabrikplanung	Ziele der Digitalen Fabrikplanung zur Unterstützung des Nutzens
Höhere Wirtschaftlichkeit der Planungslösungen	► Optimale und standardisierte Planungsprozesse ► Wiederverwendung von Best-Practise-Lösungen und Teilmodellen aus vorhergehenden Planungen
Größere Planungseffizienz und damit verkürzte Planungszeiten	► Parallelisierung von Prozessen der Produktentwicklung, Produktion, Einkauf und weiterer Unternehmensbereiche ► Einsatz digitaler und standardisierter Prozesse, Methoden und Werkzeuge
Höhere Planungsqualität	► Frühzeitiges Erkennen von Planungshürden und -fehlern ► Schnelle Analyse und Bewertung von Planungsalternativen
Bessere Beherrschung der Prozessqualität und damit der Produktqualität	► Verfügbarkeit aktueller, verifizierter und redundanzfreier Daten
Verbesserte Kommunikation und Zusammenarbeit von Produktentwicklung, Produktion, Gebäudeplanern und Lieferanten	► Gemeinsame und integrierte Datenmodelle
Nachhaltigerer Erwerb und Erhalt von Wissen	► Gemeinsame Daten- und Informationsbasis für die Planung und Dokumentation
Einfache und schnelle Zusammenarbeit in räumlich verteilten Teams	► Jederzeit und überall verfügbare Daten und Informationen
Verbesserte Entscheidungsfindung	► Visualisierung und Verfügbarkeit vollständiger Daten und Informationen für Entscheidungsprozesse
Beschleunigung des Anlaufs	► Frontloading durch Digital Mock Ups ► Simulation der Inbetriebnahme

Tabelle 1: Nutzen und Ziele der Digitalen Fabrikplanung.

die in der VDI 5200 Blatt 1 (2011) als Planungsphase bezeichnet wird. Teilweise wird für einige Planungsphasenmodelle in Abb. 2 auch deren dritte Prozessebenen dargestellt, um eine Differenzierung der Fabrikplanungssystematiken kenntlich zu machen.

Sowohl Felix (1998) als auch Wiendahl, Reichardt, Nyhuis (2014) haben Fabrikplanungssystematiken mit umfangreicher Berücksichtigung auch des Gebäudeentstehungsprozesses, d.h. des Gebäudeplanungs- und Gebäudeerstellungsprozesses, entwickelt. Basis dafür ist die HOAI (Werner, Pastor, 2017), die auch ein Vorgehensmodell für den Gebäudeentstehungsprozess beinhaltet. Damit sind die zwei größten und umfassendsten Planungsbereiche Produktion und Gebäude in Planungsphasenmodellen zur Fabrikplanung enthalten.

Dagegen ist der Aspekt der Digitalisierung nicht explizit in den Planungsphasenmodellen zur Fabrikplanung berücksichtigt. Die Digitale Fabrik wird bei Grundig (2015) nur auf einer Seite kurz erwähnt. Wiendahl, Reichardt, Nyhuis (2014) beschreiben auf 16 Seiten das Konzept der Digitalen Fabrik mit dazugehörigen digitalen Werkzeugen und Simulationsbeispielen sowie das Building Information Modeling (BIM), ohne das Thema in ihre Prozesse des Planungsphasenmodells zu integrieren.

Die Digitalisierung des Unternehmens verändert nicht nur die Unternehmensprozesse als solche, sondern auch die Planungsprozesse durch den Einsatz neuer Methoden, die durch Software oder im Kontext von Softwareanwendungen erst möglich werden. Ein Beispiel ist die Planung in verteilten Teams, die durch Application Sharing in komplexen 3D-Planungs- und Visualisierungswerkzeugen, wie z.B. visTable der Firma Plavis, deutlich vereinfacht wurde, indem räumlich verteilten Teammitglieder im virtuellen Planungsraum der virtuellen Fabrik gleichzeitig arbeiten und interagieren können.

Dementsprechend ist auch der Fabrikplanungsprozess dahingehend weiter zu entwickeln, um die Möglichkeiten der Digitalisierung bereits im Planungsprozess und nicht erst im Betrieb möglichst vollumfänglich anzuwenden und auszuschöpfen. Es ist somit ein neues bzw. erweitertes Planungsphasenmodelle für die Digitale Fabrikplanung zu entwickeln, das die Treibereffekte der Digitalisierung und die in Kapitel 1.2 dargestellten Nutzeneffekte erschließt. Im folgenden Kapitel 3 wird ein Planungsphasenmodelle für die Digitale Fabrikplanung vorgestellt, das auf den bekannten Fabrikplanungssystematiken und deren Planungsphasenmodellen aufsetzt und Elemente der Digitalen Fabrik integriert.

3 Aufbau und Umsetzung eines Planungsvorgehens für die Digitale Fabrikplanung

3.1 Vorgehensweise zur Planungsvorgehensentwicklung

Das Planungsvorgehen für die Digitale Fabrikplanung erfolgt sowohl mit einem Top-Down- als auch Bottom-Up-Ansatz.

Top-Down-Modellierung

In der Top-Down-Modellierung werden in einem ersten Schritt die existierenden

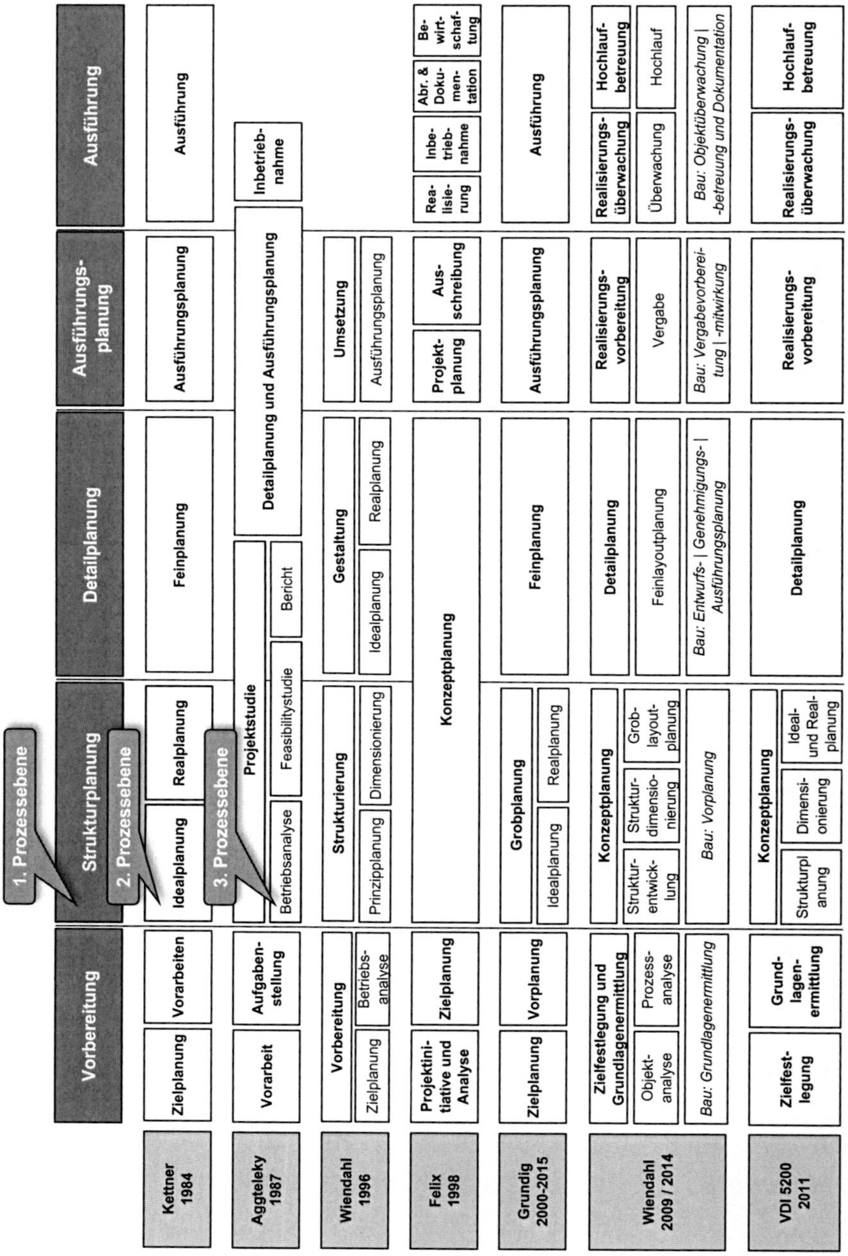

Abbildung 2: Fabrikplanungssystematiken im Vergleich. (Grundstruktur in Anlehnung an Schuh, Gottschalk, Lösch, Wesch, 2007, S. 195–199)

Fabrikplanungssystematiken zu einer umfassenden ganzheitlichen Vorgehensweise zusammengeführt. Die Fabrikplanungssystematik von Wiendahl, Reichardt, Nyhuis (2014) stellt die Hauptbasis dar, da dieses Modell am weitest gehenden ist und auch den Gebäudeentstehungsprozess umfassend berücksichtigt. Mit diesem ersten Schritt zum Aufbau des Planungsvorgehens werden alle strukturellen und inhaltlichen Belange der Fabrikplanung und Gebäudeplanung berücksichtigt.

Im zweiten Schritt werden die für die Digitalisierung erforderlichen Prozesselemente identifiziert und in das Planungsvorgehen integriert. Dazu gehören z.B. Datenstrukturanalysen und Datendigitalisierungen, um analog vorliegende Daten für die Verwendung und Verarbeitung im digitalen Planungsprozess entsprechend aufzubereiten. Auch die Datenmodellierung und die Datenintegration von Produktion und Gebäude werden durch entsprechende Prozesse abgestimmt und ganzheitlich betrachtet. Ziel ist die zur Verfügungsstellung aktueller, verifizierter und redundanzfreier Daten sowie der Aufbau und die Anwendung gemeinsamer und integrierter Datenmodelle.

Mit der Top-Down-Modellierung wird ein Rahmenwerk entwickelt und umgesetzt, das die Grundlage für die Zielerreichung der Digitalen Fabrikplanung (vgl. Kapitel 1.2) schafft.

Bottom-Up-Modellierung

Die wesentlichen Ergänzungen des Planungsvorgehens für die Digitale Fabrikplanung kommen aus der Bottom-Up-Modellierung. Aus der Literatur, Praxisberichten und Projekterfahrungen sowie aus einer dedizierten Marktrecherche zu Softwarewerkzeugen (vgl. Abb. 3) und Hardwarewerkzeugen (PC-Systeme, Powerwalls, Head Mounted Devices (HMD) u.a.) der Digitalen Fabrik und deren Zusammenspiel werden die Treiber ermittelt, die die Fabrikplanungsprozesse unterstützen oder gänzlich verändern. Der Schwerpunkt der Untersuchung sind Systeme zur Produktionsplanung und Gebäudeplanung und -erstellung. Die Systemrecherche und -analyse umfasst folgende Inhalte:

- Name und Hersteller
- Prozesse und Funktionen, die in der Fabrikplanung unterstützt werden bzw. mit dem System ausgeführt werden können
- Zuordnung zu Prozessen bzw. Aktivitäten im Planungsvorgehensmodell
- Bewertung des Systems hinsichtlich Prozessunterstützung, -erweiterung, -veränderung oder Möglichkeit eines neuen Prozesses

Die Prozesse und Aktivitäten des Planungsvorgehens werden den Analyseergebnissen entsprechend ergänzt, erweitert oder neu definiert. Insbesondere die Neudefinition von Prozessen und Aktivitäten, die ein effizienteres und effektiveres Planen ermöglichen, verändern die Struktur und Inhalte auf den tiefer liegenden Prozessebenen teilweise erheblich. Ein signifikantes Beispiel dafür ist das Virtual Collaborative Engineering (VCE), das wesentlich schnellere und qualitativ bessere Prozesse in

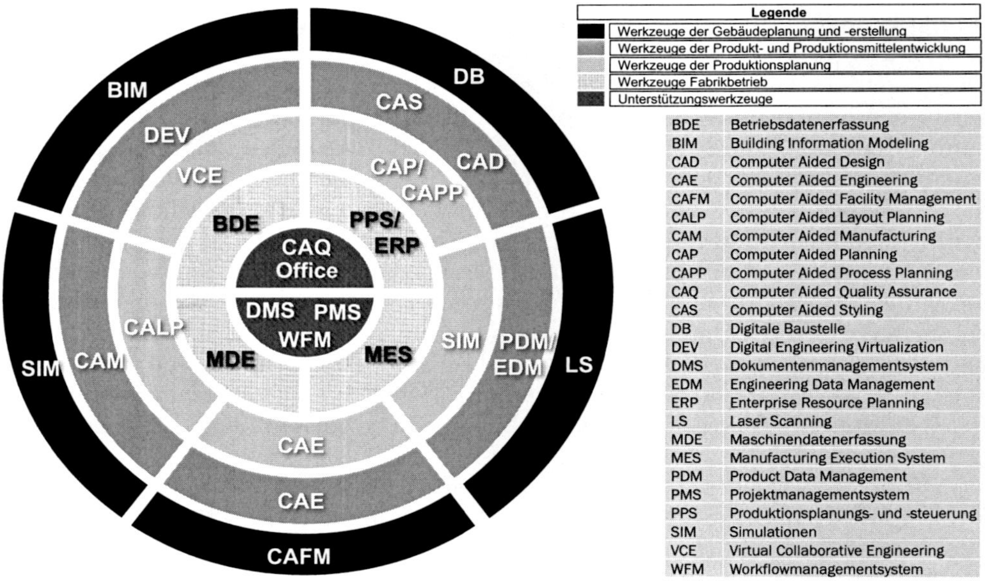

Abbildung 3: Softwaresysteme im Kontext der Digitalen Fabrik.

der globalen Zusammenarbeit von Fabrikplanungsteams ermöglicht (vgl. Augustin, 2014a - 2014d).

3.2 Planungsvorgehen für die Digitale Fabrikplanung

Das aus den in Kapitel 3.1 beschriebenen Modellierungen entwickelte Planungsvorgehen für die Digitale Fabrikplanung ist in Abb. 4 für die ersten zwei bis drei Prozessebenen dargestellt. Insgesamt umfasst das Prozessmodell des Planungsvorgehens sechs Prozessebenen. Auf der untersten Ebene sind sowohl Methoden als auch die Werkzeuge insbesondere der Digitalen Fabrik mit den durch sie unterstützten Funktionen beschrieben. Das Modell ist generisch gestaltet, so dass es die gängigen Fabrikplanungsfälle Neuplanung und Umplanung sowie bedingt die Revitalisierung unterstützt. Die Revitalisierung wird erst ab den Prozessschritten unterstützt, die nach der Verfügbarmachung folgen, d. h. die dann der Neu- oder Umplanung entsprechen. Der Rückbau ist nicht im Modell berücksichtigt.

3.3 Anwendung

Das neu entwickelte Planungsvorgehen für die Digitale Fabrikplanung unterstützt die Fabrikplaner mit dem Ziel, die in Kapitel 1.2 genannten Ziele einer modernen Fabrikplanung unter Nutzung der Technologien der Digitalen Fabrik zu erreichen. Das Pla-

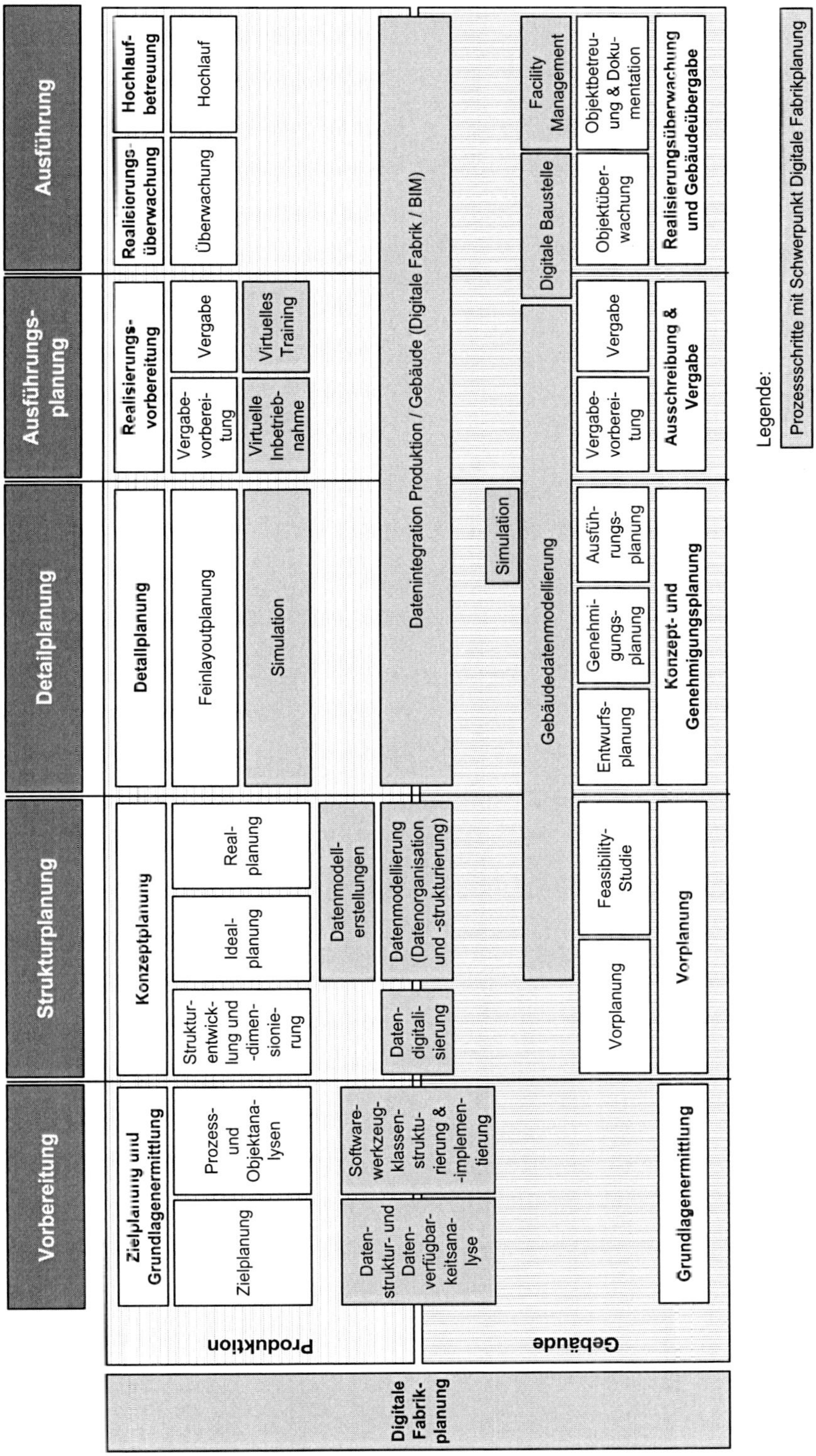

Abbildung 4: Planungsvorgehen der Digitalen Fabrikplanung.

nungsvorgehen unterstützt dabei folgende Bereiche:

- Das Prozessmodell mit seinen Subprozessen und Aktivitäten strukturiert das Vorgehen in der Digitalen Fabrikplanung und dient als Grundlage für die Erstellung von Projektplänen, da in dem Prozessmodell unmittelbar alle relevanten Aktivitäten für ein Fabrikplanungsprojekt dargestellt sind.
- Die Planung der Projekte bzgl. des Einsatzes von Soft- und Hardwaresystemen wird deutlich vereinfacht, da zu den Prozessen die passenden Systeme und deren Prozessunterstützungsfunktion dargestellt sind.
- Die Darstellung der Soft- und Hardwaresysteme und deren Verwendung in Prozessen und für Aktivitäten zeigt Potentiale auf, die Unternehmen zur Entscheidungsunterstützung für die Systemauswahl und den Systemeinsatz dienen.

4 Fazit

Das Planungsvorgehen für die Digitale Fabrikplanung bietet durch seine Erweiterung hinsichtlich der Integration der Aspekte der Digitalen Fabrik eine weitergehende Basis für das Management und die Umsetzung von Fabrikplanungsprojekten. Es zeigt die Potentiale der Digitalisierung von Planungsprozessen auf und gibt dem Fabrikplaner konkrete Unterstützung bei der Berücksichtigung neuer Technologien im Planungsprozess. Damit werden die Potentiale, die sich aus der Digitalen Fabrik für den Fabrikplanungsprozess ergeben, frühzeitig berücksichtigt und führen das Unternehmen auch in der Planung in die digitale Zukunft.

Aufgrund der sich schnell entwickelnden Technologien im Umfeld der Digitalen Fabrik, wird es erforderlich sein, die Einsatzmöglichkeiten der Soft- und Hardwaresysteme kontinuierlich nachzupflegen. Es wird auch auf absehbare Zeit immer darum gehen, die Vielfalt der Systeme in Einklang zu bringen, da ein vollständig durchgängiges System nicht existiert und in absehbarer Zukunft auch nicht verfügbar sein wird. Umso wichtiger ist es, sich anhand des entwickelten Planungsvorgehens für die Digitale Fabrikplanung klar zu machen, mit welchen Systemen, was erreichbar ist und welche Prozesse tatsächlich positiv unterstützt werden können. Dabei hilft das vorgestellte Planungsvorgehen und bietet auch für die Praxis einen hohen Mehrwert.

Literatur

Augustin, H. (Hrsg.) (2014a): Management virtueller, internationaler Teams: Vorgehensmodelle zur Kompetenzentwicklung. Aachen: Shaker.

Augustin, H. (Hrsg.) (2014b): Management virtueller, internationaler Teams: Reifegradmodelle für die Teamentwicklung. Aachen: Shaker.

Augustin, H. (Hrsg.) (2014c): Management virtueller, internationaler Teams: Leistungsmessung und -bewertung. Aachen: Shaker.

Augustin, H. (Hrsg.) (2014d): Management virtueller, internationaler Teams: Nutzen, Chancen und Risiken von Kommunikationstechnologien. Aachen: Shaker.

Bracht, U.; Geckler, D.; Wenzel, S. (2011): Digitale Fabrik: Methoden und Praxisbeispiele. Berlin Heidelberg: Springer.

Felix, H. (1998): Unternehmens- und Fabrikplanung: Planungsprozesse, Leistungen und Beziehungen. München, Wien: Carl Hanser.

Grundig, C.-G. (2015): Fabrikplanung: Planungssystematik - Methoden - Anwendungen (5. aktual. Auflage). München: Carl Hanser.

Kühn, W. (2006): Digitale Fabrik. Fabriksimulation für Produktionsplaner. München [u.a.]: Hanser.

Schenk, M./ S. Wirth/ E. Müller (2014): Fabrikplanung und Fabrikbetrieb: Methoden für die wandlungsfähige, vernetzte und ressourceneffiziente Fabrik (2. Auflage). Berlin, Heidelberg: Springer Vieweg.

Schmigalla, H. (1995): Fabrikplanung: Begriffe und Zusammenhänge. München: Carl Hanser Verlag.

Schuh, G.; Gottschalk, S.; Lösch, F.; Wesch, C. (2007): Fabrikplanung im Gegenstromverfahren. wt Werkstattstechnik online 97 (2007) 4, S. 195- 199.

VDI 4499 Blatt 1 (2008): Digitale Fabrik. Grundlagen. Berlin: Beuth.

VDI 4499 Blatt 2 (2011): Digitale Fabrik. Digitaler Fabrikbetrieb. Berlin: Beuth.

VDI 4499 Blatt 3 (2016): Digitale Fabrik. Datenmanagement und Systemarchitekturen. Entwurf. Berlin: Beuth.

VDI 4499 Blatt 4 (2015): Digitale Fabrik. Ergonomische Abbildung des Menschen in der Digitalen Fabrik. Berlin: Beuth.

VDI 5200 Blatt 1 (2011): Fabrikplanung. Planungsvorgehen. Berlin: Beuth.

VDI 5200 Blatt 2 (2016): Fabrikplanung. Morphologisches Modell der Fabrik zur Zielfestlegung. Berlin: Beuth.

VDI 5200 Blatt 3 (2016): Fabrikplanung. Phasenmodell zur Gestaltung globaler Produktionsnetze. Berlin: Beuth.

VDI 5200 Blatt 4 (2016): Fabrikplanung. Erweiterte Wirtschaftlichkeitsrechnung. Berlin: Beuth.

Werner, U.; Pastor, W. (2017): VOB - Vergabe- und Vertragsordnung für Bauleistungen. HOAI - Honorarordnung für Architekten und Ingenieure (33. Auflage). München: dtv.

Wiendahl, H.-P.; Reichardt, J.; Nyhuis, P. (2014): Handbuch Fabrikplanung: Konzept, Gestaltung und Umsetzung wandlungsfähiger Produktionsstätten (2. Auflage). München Wien: Carl Hanser.

Produktionslogistikplanung mit einer morphologiegestützten Planungssystematik

Harald Augustin[a], Natalie Drexler[b]

[a] *ESB Business School, Reutlingen University, D-72762 Reutlingen*
[b] *70469 Stuttgart*

Abstract

An die Produktionslogistik werden komplexe Anforderungen gestellt, so dass eine systematische Vorgehensweise zu deren Planung erforderlich ist. Aufgrund der Abhängigkeiten in der Supply Chain muss die Produktionslogistik immer vom Lieferanten beginnend geplant werden. Trotz der Vielfalt an Planungssystematiken bieten diese alle nur sehr wenig Unterstützung bei der Gestaltung der einzelnen Planungsobjekte in der Logistikkette an, sodass die Planung und Lösungsfindung häufig intuitiv erfolgt und sehr stark vom individuellen Erfahrungswissen des Logistikplaners abhängt. Im Folgenden wird eine Planungssystematik, die auf Morphologien basiert, vorgestellt, die eine durchgängige Planung von Produktionslogistiksystemen ermöglicht und den Planer umfangreich mit Informationen zu Gestaltungsmöglichkeiten und praktischen Umsetzungen unterstützt.

---

1 Planungssystematiken für die Produktionslogistik

1.1 Betrachtungsrahmen Produktionslogistik

Die Logistikkette eines Unternehmens reicht vom Beschaffungsmarkt bis zum Absatzmarkt und wird von Pfohl (2010) in vier Subsysteme unterteilt:

- Beschaffungslogistik,
- Produktionslogistik,
- Distributionslogistik,
- Entsorgungslogistik. (Pfohl, 2010, S. 17)

Die Trennung von Beschaffungslogistik und Produktionslogistik ist aus Sicht der Unternehmensziele hohe Teileverfügbarkeit bei niedrigen Beständen und eines ganzheitlichen und integrativen Betrachtungsansatzes des Supply Chain Managements nicht mehr sinnvoll, da die Prozesse als auch die verwendeten Technologien zunehmend integral miteinander verknüpft sind. Für die Planung und Gestaltung der Produktionsversorgung sind die Beschaffungs- und Produktionslogistik integrativ zu betrachten. Davon unbenommen, ist eine organisatorische Trennung von Unternehmensfunktionen möglich. In diesem Sinne wird im Folgenden der Begriff Produktionslogistik als umfassend für die Funktionen der Beschaffungs- und Produktionslogistik verwendet. Weiter wird die so definierte Produktionslogistik in zwei Bereiche unterteilt, um Planungsbereiche und -umfänge besser gegeneinander abgrenzen zu können (vgl. Abb. 1):

- Die Produktionsversorgungslogistik umfasst die Materialversorgung der Produktion, d.h. sie bildet die Logistikkette vom Beschaffungsmarkt bis an den Arbeitsplatz in der Produktion ab.
- Die Produktionsentsorgungslogistik umfasst die Verteilung von fertiggestellten Teilen bzw. Gütern in andere Produktionsbereiche als auch die Verbringung in das Fertigwarenlager bzw. Distributionslager sowie die Müllentsorgung.

1.2 Planungssystematiken

Für die Planung der Produktionslogistik finden sich unterschiedliche Vorgehensmodelle in der Literatur. Abb. 2 zeigt die Unterschiede in Aufbau und Aufteilung der Planungsinhalte der verschiedenen Planungssystematiken. Einige Planungsverfahren sind grober gehalten als andere und fassen in einer Phase mehrere Planungsinhalte zusammen. So umfasst z.B. die Phase 2 der integrierten Beschaffungs- und Bereitstellungsplanung nach Pawellek (2000) die Erstellung von Ideal- und Realstrukturen, wohingegen die Materialflussplanung nach Arnold und Furmans (2009) diese Planungsinhalte auf zwei Phasen aufteilt. Ausgangspunkt der Planung stellt bei allen Verfahren die Analyse des Ist-Zustands dar, um die Planungsdaten sowie die bestehenden Randbedingungen aufzunehmen und daraus die Planungsziele abzuleiten. Diese Planungsaufgabe ist innerhalb der betrachteten Verfahren entweder auf zwei Phasen aufgeteilt (z.B. bei Jünemann (1989)) oder beschränkt sich auf eine Phase. In Abb. 2 ist dieser einheitliche Ausgangspunkt grau hinterlegt.

Die in Abb. 2 dargestellten Planungssystematiken bilden die Basis für die Entwicklung der neuen, Morphologie gestützten Planungssystematik, die in Kapitel 2 vorgestellt wird.

2 Produktionslogistikprozessreferenzmodell

Referenzmodelle sind übergeordnete allgemeingültige Modelle und dienen als Ausgangspunkt für die Entwicklung anwendungsbezogener Modelle (vgl. Hars, 1994,

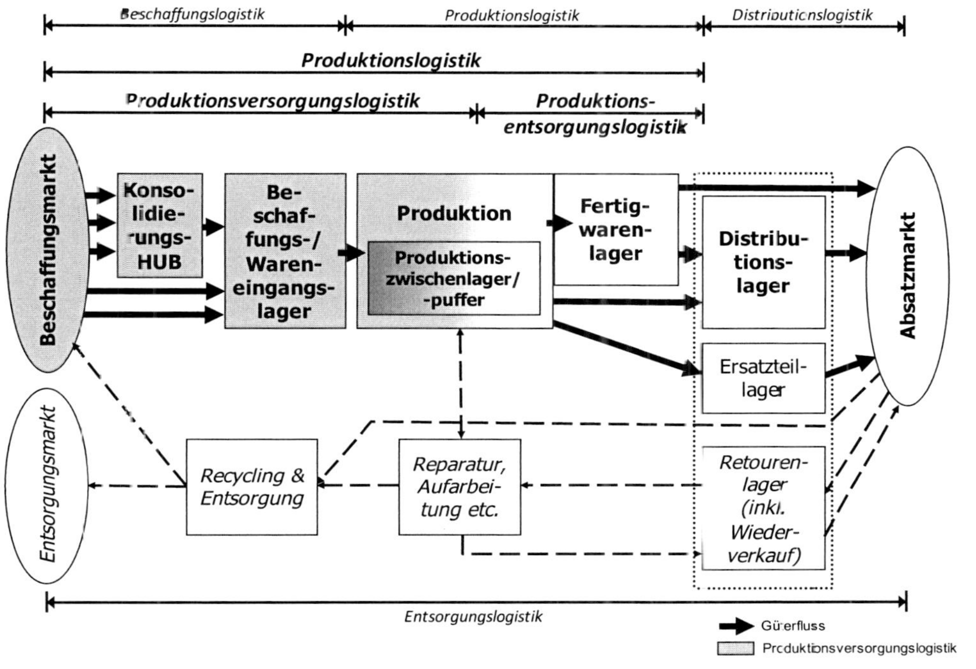

Abbildung 1: Die Produktionslogistik innerhalb der Unternehmenslogistikkette.

S. 15). Da sie eine problemübergreifende Modellierung darstellen, sind sie allgemeiner und umfassender als spezielle Modelle (vgl. Corsten und Gössinger, 2008, S. 134). Sie erleichtern und beschleunigen den Prozess der spezifischen Modellentwicklung, da auf bereits identifizierte Strukturen und Prozesse zurückgegriffen werden kann (vgl. Schneider, 2016, S. 82).

Ein Beispiel eines Prozessreferenzmodells in der Logistik ist das „Supply Chain Operations Reference Modell" (vgl. Colsman, 2002, S. 32). Es bildet die logistische Kette von der Beschaffung bis zur Distribution einschließlich der Retouren ab (vgl. Schulte, 2017, S. 783). Aufgrund des zu hohen Abstraktionsgrads eignet es sich jedoch nicht, um als Prozessreferenzmodell für das Produktionslogistiksystem genutzt zu werden.

Für die Entwicklung einer Planungssystematik ist es erforderlich, den Produktionslogistikprozess detaillierter in einem Produktionslogistikprozessreferenzmodell zu erfassen, um darauf basierend die Planungsschritte und damit die Planungssystematik aufzubauen. In Abb. 3 ist das Produktionslogistikprozessreferenzmodell für den Hauptprozess „Produktionsversorgungslogistik" (vgl. Kapitel 1.1) dargestellt. Die im Prozessreferenzmodell in Kreisen hinterlegten Nummern stellen den Bezug zu den Planungsschritten der Planungssystematik her (vgl. Abb. 6).

Die Abb. 4 zeigt die oberste Prozessebene des ProduktionslogistiksystemDie wei-

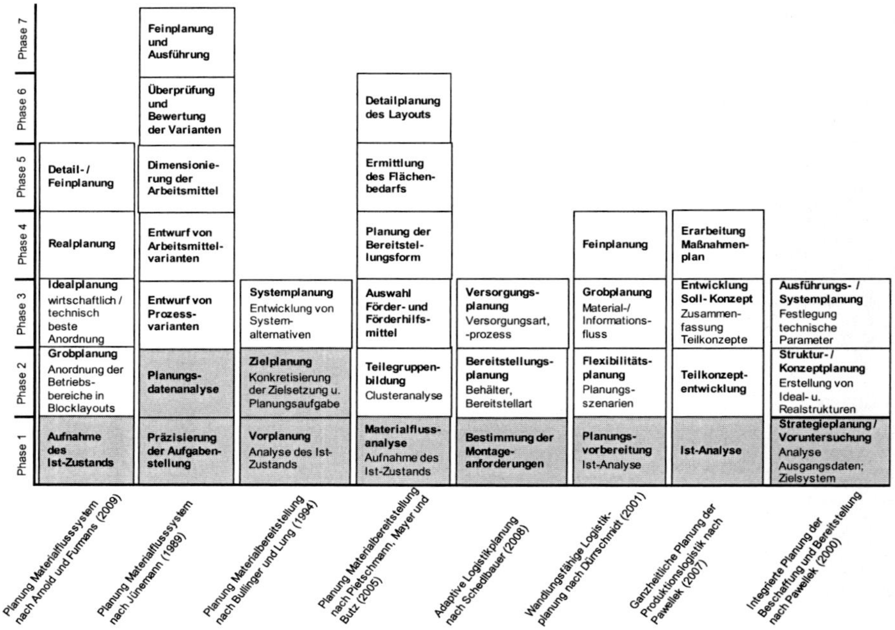

Abbildung 2: Übersicht zu Phasen verschiedener Planungssystematiken für die Produktionslogistik.

tere Unterleitung der Prozesse in Teilprozesse auf den unteren Ebene der Prozesshierarchie erfolgt in den Morphologien (vgl. Abschnitt 3), um eine bessere Übersichtlichkeit und einfachere Planungsanwendung zu erreichen.

Das Prozessreferenzmodell umfasst die maximale Ausprägung der Prozesskette, d.h. die maximal anfallenden Prozesse. Je nach Planungsfall müssen nicht immer alle Prozesse relevant sein, sodass ggf. einzelne übersprungen oder ausgelassen werden können. Darüber hinaus nimmt das gewählte Anlieferkonzept durch Bestimmung der Senke der außerbetrieblichen Logistikprozesse erheblichen Einfluss auf den Umfang der zu planenden innerbetrieblichen Teilprozesse. Abb. 4 zeigt mögliche Prozessketten für die Produktionslogistik ab dem Wareneingang. Im Falle einer handhabungsarmen Versorgung kann z.B. eine einstufige Materialbereitstellung, d.h. Direktbelieferung eines dezentralen Supermarkts ab Wareneingang oder ab Lieferant, dargestellt werden (vgl. Abb. 4, Nr. 3).

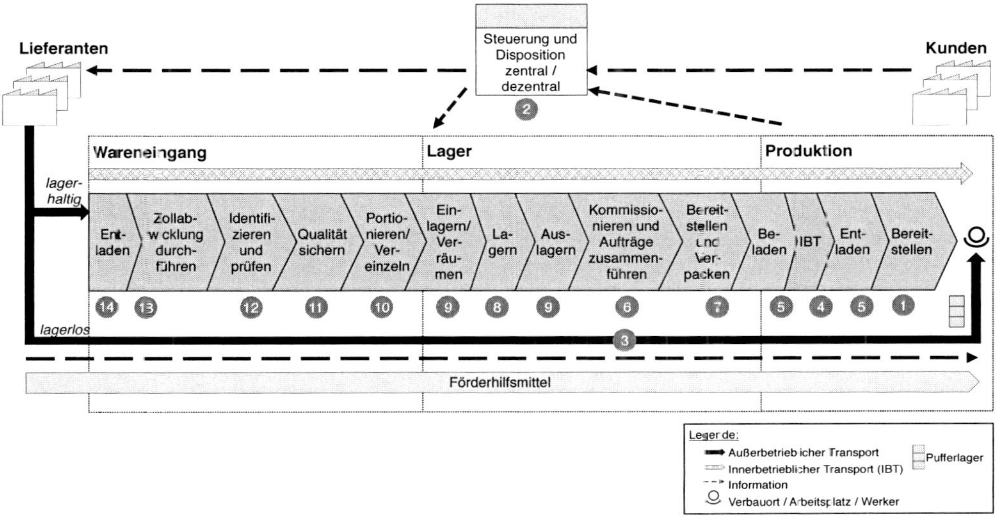

Abbildung 3: Prozessreferenzmodell der Produktionslogistik.

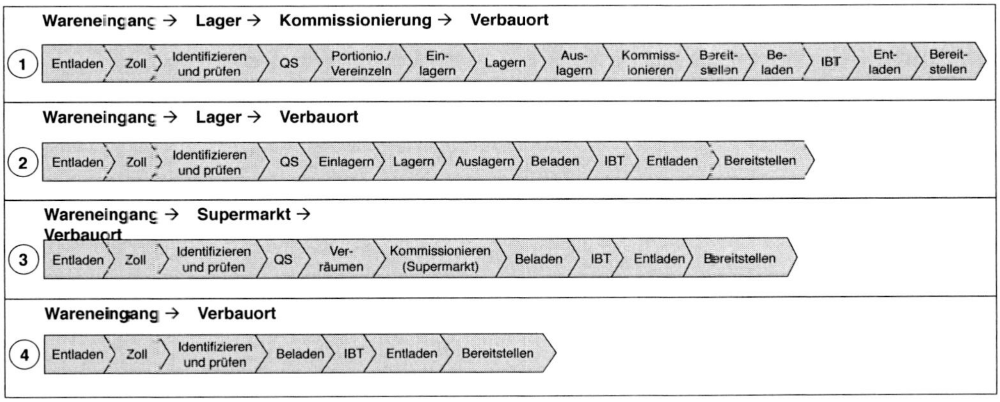

Abbildung 4: Verschiedene Prozessketten der Produktionslogistik.

3 Morphologien für die Produktionslogistikgestaltung

3.1 Morphologische Methode

Die Methode des morphologischen Kastens wurde von dem Schweizer Astrophysiker Fritz Zwicky im Jahr 1956 entwickelt (vgl. Baum, 2011, S. 25). In der Literatur ist diese Methode auch unter den Bezeichnungen „Morphologisches Schema“ oder nur „Morphologie“ bekannt (vgl. Ponn und Lindemann, 2011, S. 116–117). Als gängigste morphologische Methode zielt sie darauf ab, eine Problemstellung oder einen

Sachbereich durch systematische Zerlegung lückenlos und ohne Überschneidungen zu strukturieren und gesamtheitlich zu erfassen (vgl. Disselkamp, 2012, S. 161). „Die Anwendung der morphologischen Methode gibt uns die größtmögliche Sicherheit, dass nichts vergessen wird, was für die Beleuchtung aller Aspekte eines vorgegebenen Problems von Wichtigkeit ist" (Zwicky, 1966, S. 114). Aus diesem Grund liegen die Stärken dieser Methode insbesondere in ihrem Informationsgehalt sowie Detaillierungsgrad (vgl. Disselkamp 2012, S. 161).

Komplexe Aufgabenstellungen und Sachverhalte werden anhand ausgewählter Parameter bzw. Merkmale in Teilaspekte zerlegt (vgl. Gessler, 2014, S. 110). Nach erfolgter Merkmalsbildung werden alle Ausprägungen bzw. Gestaltungsmöglichkeiten der Merkmale nebeneinander in das eindimensionale Ordnungsschema des morphologischen Kastens übertragen (vgl. Gessler, 2014, S. 110.). Durch die tabellenartige Struktur ordnet der morphologische Kasten jedem Merkmal die möglichen Teillösungsideen bzw. Gestaltungsmöglichkeiten zu (vgl. Ponn und Lindemann, 2011, S. 116–117). Zur Erleichterung der späteren Entwicklung von Gesamtlösungskonzepten empfehlen Ponn und Lindemann die Merkmale entsprechend ihrer Lösungsdominanz in der Spalte des morphologischen Kastens anzuordnen (vgl. Ponn und Lindemann, 2011, S. 102).

3.2 Aufbau der Morphologien für die Planungssystematik

Der Aufbau der Morphologie für die Planungssystematik orientiert sich vollständig an dem in Abschnitt 3.1 beschriebenen Verfahren. Als Quellen für den inhaltlichen Aufbau der Morphologien dienen Literaturquellen sowie Normen und Richtlinien. Für jeden Prozess des Prozessreferenzmodells (vgl. Abb. 3) liegt je nach dessen Komplexität mindestens eine, in vielen Fällen mehrere Morphologien vor. Tab. 1 zeigt exemplarisch einen Auszug aus der Morphologie für den Prozess „Entladen", der in mehrere Teilprozesse aufgeteilt ist. Für jeden Teilprozess werden die Merkmale mit ihren möglichen Ausprägungen dargestellt. Beispiele für einzelne Ausprägungen geben dem Planer Hinweise auf konkrete Realisierungsmöglichkeiten. Die Quellenangaben sind insbesondere dann von Bedeutung, wenn auf Normen oder Richtlinien verwiesen wird. In diesen können weitere Details zu Gestaltungsmöglichkeiten, die manchmal auch in Listen- oder Morphologieform vorliegen, recherchiert werden.

4 Morphologie gestützte Planungssystematik

4.1 Planungsvorgehen nach dem Line-Back-Prinzip

Dem Vorgehen in der Planungssystematik wird das „Line-back"-Planungsprinzip aus der adaptiven Logistikplanung (Schedlbauer, 2007) zugrunde gelegt. Demnach wird die Logistikkette entgegengesetzt des physischen Materialflusses ausgehend von dem Verbau- bzw. Bereitstellort in der Produktion geplant (vgl. Abb. 5). Dies impliziert, dass der Bereitstellort den Engpass im System darstellt, da die Logistikflächen umso knapper und wertvoller werden, je näher sie sich am Produkt befinden (vgl. Klug,

Ausführungsart	Merkmal	Ausprägungen				Beispiele	Quellen
Art der Entladung	Art der Entladung	1) Heck-entladung	2) Seitenentladung (ein-/beidseitig)	3) Dach-entladung			Martin 2014: 303; Hompel / Schmidt / Nagel 2007: 312 Miebach / Schlichting 2007: 138-139
		1) Teil-entladung	2) Komplett-entladung			2) Beladener Routenzug fährt aus LKW-FTS automatische LKW-Entladung (s.u.)	
	Entladung über Rampe	1) ja	2) nein			1) Seiten-, Kopf-, Dockrampe, Rampe in Sägezahnform	Schulte 2017: 228

Teilprozess	Merkmal	Ausprägungen				Beispiele	Quellen
Last aufnehmen	Automatisierungsgrad	1) manuell	2) mechanisiert (mit personengesteuertem Hilfsmittel)	3) automatisch		1) Mensch (ohne Hilfsmittel) 2) Mensch steuert Krane, Stapler, Gabelhubwagen 3) Maschine arbeitet selbsttätig; Mensch hat Kontrollfunktion	Martin 2014: 301 VDI 4420 1996: 2
				1) mit auf der Ladefläche installierter Fördertechnik	2) ohne Fördertechnik auf der Ladefläche	1) Tragketten-/ Rollenförderer, Pendelboden (Walking Floor), hängende Verladung 2) Hubkettenförderer, Gleitschienenförderer, Entladung mithilfe von FTS	
	Lastaufnahme- / -abgabemittel	1) Mensch	2) Kran	3) Hubwagen	4) Gabelstapler		Martin 2014: 313 Günthner / Schmalzl / Seemüller o. J.: 1
Lagergut von Lade- zu Pufferfläche transportieren	Quelle-Senke-Beziehung	1) 1:1	2) 1:n				
	Anzahl der pro Arbeitsspiel umgeschlagenen Ladeeinheiten	1) 1	2) >1				Hompel / Schmidt / Nagel 2007: 312
	Transportmittel						

Tabelle 1: Morphologie „Entladen“.

2010, S. 80). Da der Verbauort die Durchsatzleistung und Durchlaufzeit des Logistiksystems vorgibt, muss sich die gesamte Planung auf den Engpassbereich beziehen, was die retrograde Planungsweise erforderlich macht. Zudem ist dieses Planungsprinzip zu bevorzugen, da es den Prinzipien der Lean Philosophie entspricht, wonach alle Tätigkeiten nach den Anforderungen des Kunden, im Fall der Produktionslogistikplanung, dem Produktionsmitarbeiter ausgerichtet werden.

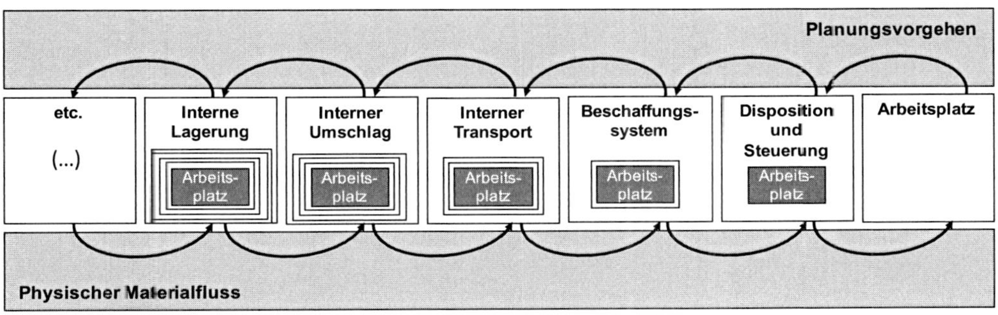

Abbildung 5: Planung nach dem „Line-back“-Prinzip (vgl. Schulte, 2017, S. 407)

4.2 Morphologie gestützte Planungssystematik

Die Planungssystematik gliedert sich in sechs Phasen:

- Phase 1: Analyse der Ausgangssituation
- Phase 2: Planung der Bereitstellung
- Phase 3: Planung der Steuerung und Disposition
- Phase 4: Planung des Beschaffungssystems
- Phase 5: Grobplanung des Material- und Informationsflusses
- Phase 6: Feinplanung

Jede Phase bündelt jeweils eine Sequenz an Planungsschritten. Die Phasen orientieren sich an den in der Literatur vorliegenden Planungssystematiken (vgl. Kapitel 1.2). Dies ermöglicht eine einfache Orientierung und Verknüpfung mit parallelen Planungsabläufen, wie z.B. der Produktionsplanung.

Das Vorgehensmodell der Planungssystematik ist auszugsweise in Abb. 6 dargestellt. Zu jedem Planungsschritt werden die relevanten Inputdaten dargestellt, die durch die planerische Aktivität die gewünschten Outputdaten liefern. Neben den Daten sind in der Planungssystematik erforderliche Dokumente und Methoden angegeben. Außerdem sind die Morphologien den entsprechenden Planungsschritten (durch die nummerierten Kreise dargestellt) zugeordnet und zeigen dem Logistikplaner die unterschiedlichen Gestaltungsmöglichkeiten zu den einzelnen Planungsparametern auf.

Die sequentiell angeordneten Planungsschritte beeinflussen sich gegenseitig, sodass bei der Planung kein strikt lineares Abarbeiten der Planungsschritte möglich ist. Neue Erkenntnisse im Planungsvorgehen können dazu führen, dass an bereits bearbeitete Schritte zurückgesprungen werden muss und diese nochmals überdacht oder ggf. neu geplant werden müssen.

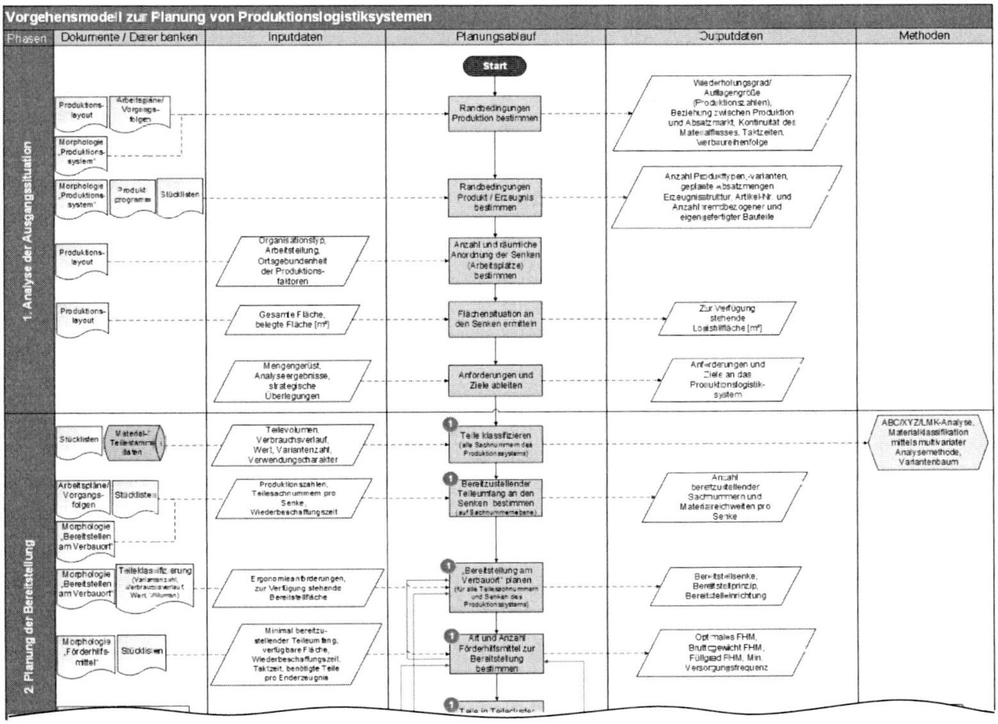

Abbildung 6: Darstellung des Vorgehensmodells der Planungssystematik zur Produktionslogistikplanung (Auszug).

5 Fazit

Die vorgestellte morphologiegestützte Planungssystematik unterstützt den Planer bei seiner Arbeit in mehrfacher Hinsicht:

- Die einfach anzuwendende Planungsvorgehensweise führt ihn nach dem Line-Back-Prinzip durch die zu planenden Prozessschritte der Produktionslogistik.
- Die in die Vorgehensweise integrierten Morphologien unterstützen ihn bei der Analyse und Definition von Lösungsräumen sowie mit konkreten Auslegungs- und Gestaltungsmöglichkeiten für die zu planenden Prozesse.

Durch Anwendung der Planungssystematik wird die Planung beschleunigt und die Ergebnisse werden valider, da alle möglichen Ausprägungen der Gestaltung bedacht und abgewogen werden. Dabei kommt es nicht mehr allein auf den Erfahrungsschatz des Planers an, da die Morphologien das Wissen aus den einschlägigen Wissensquellen zur Produktionslogistikplanung enthalten.

Die Weiterentwicklung der Planungssystematik wird der weitere Ausbau der Morphologien sein. Insbesondere die vollständige Integration von Checklisten, Morphologien etc. aus relevanten VDI-Richtlinien und Normen anstelle des aktuellen Verweises darauf wird die Anwendung der Planungssystematik noch weiter vereinfachen, da nicht parallel in Richtlinien und Normen recherchiert werden muss.

Literatur

Arnold, D.; Furmans, K. (2009): Materialfluss in Logistiksystemen (6. Auflage). Heidelberg u.a.: Springer.

Baum, H. (2011): Morphologie der Kooperation als Grundlage für das Konzept der Zwei-Ebenen-Kooperation. Zugl.: Chemnitz, Techn. Univ., Diss., 2008. Wiesbaden: Gabler.

Bullinger, H.-J.; Lung, M. M. (1994): Planung der Materialbereitstellung in der Montage. Stuttgart: Teubner.

Colsman, R. (2002): Prozessorientierung als Qualitätsprinzip der Logistik. In: Wiendahl, H.-P. (Hg.): Erfolgsfaktor Logistikqualität. Vorgehen, Methoden und Werkzeuge zur Verbesserung der Logistikleistung. 2. Auflage Berlin: Springer. S. 31–35.

Corsten, H.; Gössinger, R. (2008): Einführung in das Supply Chain Management (2. Auflage). München: Oldenbourg.

Disselkamp, M. (2012): Innovationsmanagement. Instrumente und Methoden zur Umsetzung im Unternehmen (2. Auflage). Wiesbaden: Springer.

Dürrschmidt, S. (2001): Planung und Betrieb wandlungsfähiger Logistiksysteme in der variantenreichen Serienproduktion. Zugl.: München, Techn. Univ., Diss., 2001.

Gessler, R. (2014): Entwicklung eingebetteter Systeme. Vergleich von Entwicklungsprozessen für FPGA- und Mikroprozessor-Systeme Entwurf auf Systemebene. Wiesbaden: Springer Vieweg.

Hars, A. (1994): Referenzdatenmodelle. Grundlagen effizienter Datenmodellierung. Zugl.: Saarbrücken, Univ., Diss., 1993. Wiesbaden: Gabler.

Jünemann, R. (1989): Materialfluss und Logistik. Systemtechnische Grundlagen mit Praxisbeispielen. Berlin: Springer.

Klug, F. (Hg.) (2010): Logistikmanagement in der Automobilindustrie. Berlin, Heidelberg: Springer.

Pawellek, G. (2000): Integrierte Beschaffungs- und Bereitstellungslogistik in Produktionsnetzen. In: Beschaffung aktuell (11), S. 36–38.

Pawellek, G. (2007): Produktionslogistik. Planung, Steuerung, Controlling. München: Carl Hanser.

Ponn, J.; Lindemann, U. (2011): Konzeptentwicklung und Gestaltung technischer Produkte. Systematisch von Anforderungen zu Konzepten und Gestaltlösungen (2. Auflage). Berlin, Heidelberg: Springer.

Pietschmann, N.; Mayer, A.; Butz, C. (2005): Materialbereitstellung in der Montage. Eine Planungssystematik. In: PPS Management (2), S. 23–26.

Schedlbauer, M. J. (2008): Adaptive Logistikplanung auf Basis eines standardisierten, prozessorientierten Bausteinkonzepts. Zugl.: München, Techn. Univ., Diss., Garching: Lehrstuhl für Fördertechnik Materialfluss Logistik.

Schneider, M. (2016): Lean Factory Design. Gestaltungsprinzipien für die perfekte Produktion und Logistik. München: Hanser.

Schulte, C. (2017): Logistik. Wege zur Optimierung der Supply Chain. 7. Auflage. München: Franz Vahlen.

Zwicky, F. (1966): Entdecken, Erfinden, Forschen im Morphologischen Weltbild. München: Droemersche Verlagsanstalt.

Der Einfluss der Digitalisierung auf zukünftige Unternehmens- und Produktionsstrukturen

Manfred Estler[a], Stephan Höfer[a]

[a] ESB Business School, Reutlingen University, D-72762 Reutlingen

Abstract

Es wird erwartet, dass die neuen Technologien rund um die Digitalisierung von Gesellschaft und Geschäftswelt zu revolutionären Veränderungen führen werden. Im Zusammenhang mit produzierenden Unternehmen ist hier von einer möglichen vierten Revolution unter dem Stichwort „Industrie 4.0" die Rede. Eine Frage, die damit aber unmittelbar einhergeht, ist, ob sich infolge dieser Revolution auch Organisations- und Produktionsstrukturen von Unternehmen nicht ebenfalls revolutionär ändern müssen.
Dieser Beitrag geht dieser Frage nach, indem er den momentanen Stand der wissenschaftlichen Diskussion zusammenfasst und anschließend bewertet.

1 Produktionsparadigmen in ihrer historischen Entwicklung

Neue Technologien waren stets Treiber für die Entwicklung neuer Geschäftsmodelle. Im Zusammenspiel mit der Globalisierung schaffen neue Informationstechnologien und die damit verbundene Digitalisierung von Gesellschaft und Geschäftswelt Voraussetzungen für eine mögliche vierte industrielle Revolution (Industrie 4.0). Neue Geschäftsmodelle bedingen jedoch unmittelbar neue Gestaltungsprinzipien für die Fertigung in produzierenden Unternehmen. Koren (2010) beschreibt anhand Abbildung 1 die historische Entwicklung von Produktionsparadigmen.

Die handwerkliche Fertigung des 19. Jahrhunderts war geprägt von einer hohen Produktvielfalt bei gleichzeitig geringen Stückzahlen je Produktvariante. Um Produkte für größere Käuferschichten erschwinglich zu machen, entwickelte sich zu Beginn des 20. Jahrhunderts die Massenfertigung. Sie war geprägt von einer geringen Produktvariation und hohen Stückzahlen. Die Arbeiten von Frederick Winslow Taylor (1856–1915) bildeten die wissenschaftliche Grundlage für diese Entwicklung. Der nach ihm benannte Taylorismus führte zu einer starken Arbeitsteilung in Unternehmen, was mit enormen Produktivitätssteigerungen verbunden war. Was zunächst auf

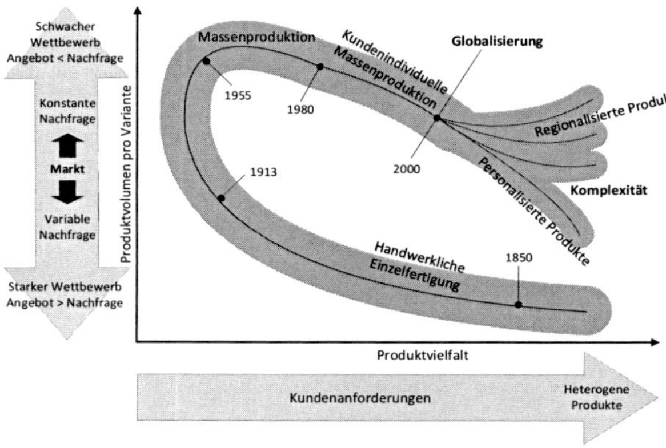

Abbildung 1: Historische Entwicklung der Produktionsparadigmen (in Anlehnung an Koren (2010))

die Produktion beschränkt war, setzte sich in den Aufbauorganisationen von Unternehmen fort. Im Zuge der Einführung der Massenfertigung bildeten sich in Unternehmen stark arbeitsteilig geprägte, tayloristische Strukturen aus. Die darauf resultierende Funktionsorientierung prägt bis heute die Struktur und Aufbauorganisation der überwiegenden Mehrzahl von Unternehmen.

Die kundenseitige Nachfrage nach individuell konfigurierbaren Produkten führt zum Ende des 20. Jahrhunderts hin zum Produktionsprinzip der Mass Customization. Diese erlaubt gegenüber der reinen Massenfertigung eine größere Anzahl von Produktvarianten derselben Produktfamilie, um die gestiegenen Kundenanforderungen besser erfüllen zu können.

Durch die Globalisierung ist der Bedarf nach kundenindividuellen Produkten einerseits, regionalisierten Produkten andererseits seit der Jahrtausendwende enorm gestiegen. Dieser Trend geht schon heute teilweise hin zur Produktion von Losgröße 1, also im Prinzip zurück zur Einzelfertigung früherer Jahre. Allerdings ist das Innovationstempo verglichen mit dieser früheren Zeit dramatisch angestiegen. Zudem hat diese kundenindividuelle bzw. personalisierte Produktion nun den Anspruch, diese neue Art der Produktion zu Herstellkosten nahe den Kosten der Massenfertigung zu bewerkstelligen. Die neuen Technologien der Digitalisierung fungieren hier als Enabler, um dieses Ziel erreichen zu können.

Diese Entwicklung wurde begleitet von bisher drei industriellen Revolutionen. Die erste industrielle Revolution um das Jahr 1750 war getrieben durch die Erfindung und den industriellen Einsatz der Dampfmaschine zur Energieerzeugung und die Mechanisierung von Handarbeit durch Maschinen. Die Entwicklung der arbeitsteiligen Massenproduktion unter Einsatz elektrischer Energie stand im Mittelpunkt der zweiten industriellen Revolution Ende des 19. Jahrhunderts. Die zunehmende Automatisierung der Produktion Mitte den 20. Jahrhunderts auf Basis neuer Informations- und Kommunikationstechnologie des beginnenden Computerzeitalters ermöglichte schließlich die dritte industrielle Revolution. Nun wird erwartet, dass neue Informationstechnologien rund um cyber-physische Systeme (CPS), Augmented Reality, Big Data, Cloud Computing und das Internet of things der Startpunkt einer neuen, der vierten industriellen Revolution sind.

Die Frage, die sich nun an dieser Stelle ergibt, ist, ob die Fähigkeit, kundenindividuelle Produkte in kleinen Losgrößen zu wettbewerbsfähigen Preisen herzustellen, ein überwiegend technologisches Problem darstellt. So beschreibt z.B. Bauernhansl (2014), wie Technologien der Informations- und Kommunikationstechnik Einfluss auf Produktentwicklung, Produktionsprozessgestaltung, Produktionsplanung und Produktionssteuerung haben werden. Im Zentrum stehen dabei cyber-physische Systeme, die die Basis der Smarten Fabrik der Zukunft darstellen. Im Zentrum der technologischen Entwicklung stehen dabei die Leistungssteigerung von Fabriken durch Vernetzung, Dezentralisierung und Autonomie der einzelnen Komponenten (Bauernhansl, 2014, S. 18). Dies wird nach Auffassung von Bauernhansl zu neuen Vorgehensweisen sowohl in der Fabrik-, wie auch der Produktionsplanung führen (Bauernhansl, 2014, S. 19f.). Die Diskussion über das Ausmaß der Veränderung bleibt aber im Wesentlichen auf die mit der Produktion befassten Bereiche von Unternehmen beschränkt. Mögliche Auswirkungen auf die Organisationsentwicklung von Unternehmen werden an dieser Stelle nicht diskutiert.

In einer von der Impuls-Stiftung des Verbandes Deutscher Maschinen- und Anlagenbau geförderten Studie zum Status der Umsetzung von Industrie 4.0 setzen sich auch Lichtblau et al. (2015) überwiegend mit technologischen Erfolgsfaktoren im Zuge der Einführung und Nutzung der neuen Informations- und Kommunikationstechnologien auseinander. Die Autoren der Studie adressieren zwar auch strategische und organisatorische Aspekte in diesem Kontext (Lichtblau et al, 2015, S. 30ff.). Diese Überlegungen richten ihr Augenmerk vor allem auf die Entwicklung neuer Geschäftsmodelle. Die strategische Verankerung von Industrie 4.0 wird als wesentlicher Erfolgsfaktor gesehen, wobei festgestellt wird, dass viele Unternehmen eben noch keinen diesbezüglichen strategischen Ansatz verfolgen. Auch in dieser Studie wird nicht untersucht, wie im Rahmen der Organisationsentwicklung das Thema der Digitalisierung in Unternehmen adressiert werden sollte.

In diesem Beitrag soll jedoch genauer betrachtet werden, ob nicht ein umfassendes oder gar radikales Change Management im Unternehmen erforderlich ist, das gerade auch die Organisationsform von Unternehmen mit einbezieht.

2 Organisationsentwicklung im Wandel der Zeit

Schon ein Blick in die Historie der Organisationsentwicklung lässt unterschiedliche Gestaltungsansätze für Unternehmen erkennen. Für Kosiol (1962) steht die Strukturgestaltung des Unternehmens im Sinne der Aufbauorganisation im Vordergrund. Erst in einem nachfolgenden Schritt werden die Abläufe innerhalb der Organisation definiert. Die Gestaltungsprämisse lautet entsprechend „process follows structure". Auf dieser Basis lässt sich das Ziel der Ressourceneffizienz, wie sie für die Massenproduktion entscheidend war, in hohem Maße erreichen. In Zeiten der Globalisierung mit gesättigten Käufermärkten und der Anforderung einer sehr starken Kundenorientierung bei gleichzeitig intensiven Wettbewerb und hohem Tempo der technologischen Entwicklung erweist sich dieses Organisationsprinzip als zunehmend hinderlich. Wird von Unternehmen hohe Flexibilität gefordert, um den hochdynamischen Marktveränderungen Rechnung zu tragen, fehlt es Unternehmen heute oft an der Schnelligkeit, um auf diese Anforderungen zu reagieren.

Einen Gegenpol zu den Arbeiten Kosiols stellen die nahezu zeitgleich erfolgten Untersuchungen und Ergebnisse von Chandler (1962) dar. Aus der Analyse der strategischen Restrukturierung von vier bedeutenden amerikanischen Großunternehmen formuliert er seine These „structure follows strategy". Eine seiner Schlussfolgerungen ist, dass die Unternehmensstrategie die Strukturgestaltung im Unternehmen festlegt. Die Unternehmensstrategie entwirft also den übergeordneten Plan, wie die Ressourcen des Unternehmens dem erwarteten Bedarf des Marktes zugeordnet werden. Damit führen veränderte Marktbedingungen zu Veränderungen der Unternehmensstrategie, was wiederum Anpassungen der Unternehmensstruktur zur Folge hat. Chandler sieht in der Unternehmensstrategie die Definition und Festlegung der langfristigen Ziele des Unternehmens. Die Unternehmensführung sollte sich stets an diesen Zielen ausrichten und das Unternehmen entsprechend gestalten.

Chandler's These „structure follows strategy" wird durchaus kontrovers diskutiert (Bach, Brehm, Buchholz und Petry, 2017, S. 14). Bach et al. stellen fest, dass eine Einschränkung aber dahingehend möglich ist, dass bestimmte Strategieoptionen für Unternehmen auf Basis der vorherrschenden organisatorischen Gegebenheiten gar nicht oder nur mit unverhältnismäßigem Aufwand umsetzbar sind. Insofern hat eine Organisationsstruktur Einfluss auf die Unternehmensstrategie. Dies darf aber nicht so verstanden werden, dass eben die Strategie an der Unternehmensorganisation ausgerichtet und von ihr bestimmt sein darf.

Bach et al. gehen vielmehr noch einen Schritt weiter als Chandler, denn sie verfolgen den Ansatz „structure follows process follows strategy" (Bach, Brehm, Buchholz und Petry, 2017, S. 111). Dieser ergibt sich aus ihren Betrachtungen zur Erarbeitung einer Wertschöpfungsarchitektur im Unternehmen. Die Wertschöpfungsarchitektur bildet dabei die konzeptionelle Brücke zwischen der Strategie und der Organisation des Unternehmens. Für Bach et al. liegen die Gründe für die Erarbeitung einer Wertschöpfungsarchitektur in den seit Ende des 20. Jahrhunderts stark im Wandel begriffenen Anforderungen an Unternehmen. Globalisierung, zunehmende Kundenorientierung, ständig steigender Wettbewerb, gesättigte Käufermärkte, eine sich immer stärker beschleunigende technologische Entwicklung und immer kürzer werdende Produktlebenszyklen fordern von den Unternehmen Schnelligkeit und hohe Flexibilität, um dieser Dynamik Rechnung tragen zu können. Zurzeit ist die Veränderungsgeschwindigkeit von Technologien deutlich größer als die Anpassungsgeschwindigkeit von Unternehmen (Bach, Brehm, Buchholz und Petry, 2017, S. 147). Diese Faktoren führen ihrer Meinung nach zur Notwendigkeit einer steigenden Prozessorientierung innerhalb der Unternehmen.

Bach et al. gehen auch in knapper Form auf die Auswirkungen der digitalen Transformation auf die Organisationsentwicklung in Unternehmen ein. Sie kommen trotz des Einflusses der Digitalisierung zum Schluss, dass die Grundlagen der Organisationslehre unverändert gelten, wenngleich der Einfluss unbestritten ist. Die stets erforderliche Strukturierung der Wertschöpfungsaktivitäten sowie die Definition von Zuständigkeiten und Verantwortlichkeiten ist ihrer Meinung nach im Grundsatz ebenso zu leisten wie zuvor. Gerade für sogenannte Orchestratoren ergeben sich durch die Digitalisierung vielfältige Möglichkeiten.

Als Orchestratoren werden Wertschöpfungsmodelle bezeichnet, die sich auf einzelne Stufen der Wertschöpfungskette konzentrieren, durch die geschickte Koordination der anderen Wertschöpfungsstufen einen Mehrwert generieren und die Wertschöpfungskette kontrollieren (Bieger und Reinhold, 2011, S. 39). Bekannte Vertreter dieses Ansatzes sind z.B. die Plattformen Airbnb und Uber.

Weitere Veränderungen durch Digitalisierung, Big Data oder Internet der Dinge ergeben sich durch eine nahezu vollständige Vernetzung von Arbeitsprozessen (Bach, Brehm, Buchholz und Petry, 2017, S. 148). Dies wird nicht ohne Folgen für die Gestaltung der Arbeits- und Geschäftsprozesse bleiben können. Zudem eröffnen sich dadurch große Optimierungspotentiale für Unternehmen. Insgesamt wird die Bedeutung der Informationsflüsse gegenüber den Materialflüssen zunehmen, wobei das Internet der Dinge Wege eröffnen wird, dass materielle Objekte selbst zu Informationsträgern werden (Bach, Brehm, Buchholz und Petry, 2017, S. 155).

Es lässt sich ebenfalls feststellen, dass die Anforderungen an Wertschöpfungsprozesse hinsichtlich Agilität und Flexibilität zunehmen werden (Bach, Brehm, Buchholz und Petry, 2017, S. 187f.). Dies erfordert sicherlich in höheres Maß an Prozessorientierung in Unternehmen. Es wird aber derzeit in der Praxis nicht beobachtet, dass dies zur Auflösung oder Abflachung von Hierarchien führt, wenngleich Tendenzen zu z.B. Entscheidungsdezentralisation, Delegation und Partizipation zu erkennen sind (Bach, Brehm, Buchholz und Petry, 2017, S. 275). Die Notwendigkeit, in einem dynamischen und hochdigitalisierten Umfeld sowohl Strategie als auch noch viel öfter die Organisationsform des Unternehmens zu hinterfragen, anzupassen und zu erneuern, erscheint jedoch klar (Bach, Brehm, Buchholz und Petry, 2017, S. 378).

Sehr ausführlich und schon seit dem Erscheinen der 1. Auflage ihres Standardwerkes „Geschäftsprozessmanagement in der Praxis" im Jahr 2000 setzen sich Schmelzer und Sesselmann (2013) mit der Frage der Integration der Geschäftsprozesse in die Aufbauorganisation auseinander. Zur Befriedigung der immer weiter steigenden Kundenanforderungen propagieren sie ein Geschäftsprozessmanagement, das Organisationen befähigt, „flexibel auf Veränderungen zu reagieren und erforderliche Anpassungen vorzunehmen. Es leistet wichtige Beiträge zur Steigerung der Organisationskompetenz und -effizienz" (Schmelzer und Sesselmann, 2013, S. 2). Aus den spezifischen Merkmalen von Funktions- bzw. Prozessorganisationen (siehe Tabelle 1) ergeben sich jeweils spezifische Vorteile dieser beiden Organisationsformen. Nach Auffassung von Schmelzer und Sesselmann (2013, S. 210) geht es darum, die Geschäftsprozesse gezielter in die Aufbauorganisation zu integrieren. Sie streben keinen vollständigen Ersatz der Funktionsorganisation durch eine Prozessorganisation an, sondern schlagen Wege zu prozessorientierten Aufbauorganisationen vor.

Letztendlich geht es ihnen darum, die Geschäftsprozesse in Unternehmen gegenüber der klassischen funktionsorientierten Aufbauorganisation viel stärker in den Fokus der Organisationsgestaltung zu rücken. Sehr breit diskutieren sie die Integration

Funktionsorganisation	Prozessorganisation
Ziele: Ressourcen-, Kosteneffizienz	Ziele: Kundenzufriedenheit, Prozesseffizienz
vertikale Ausrichtung	horizontale Ausrichtung
abteilungs-, stellenbezogen	abteilungsübergreifend, prozessbezogen
Verrichtungsorientierung	Geschäftsfall-, Objektorientierung
Sozialisierung, Arbeitsteilung	Aufgabenintegration
Ressort-, Statusdenken	unternehmerisches Erfolgsdenken
Machtorientierung	Kunden- und Teamorientierung
tiefe Hierarchie, viele Schnittstellen	flache Strukturen
hoher Koordinations- und Steuerungsbedarf	Selbstabstimmung und -steuerung
zentrales Fremdcontrolling	dezentrales Selbstcontrolling
kontrollierte Informationen	freie und offene Informationen
Rationalisierungsprojekte	kontinuierliche Verbesserung
Ersatzprozesse, Redundanz	Konzentration auf Wertschöpfung
Komplexität	Transparenz

Tabelle 1: Merkmale von Funktions- und Prozessorganisationen (Schmelzer und Sesselmann, 2013

der Geschäftsprozesse in die Aufbauorganisation (Schmelzer und Sesselmann, 2013, S. 204ff.). Dabei stellen sie vor allem fest, dass es prozessorientierten Organisationen möglich ist, „flexibel auf Bedürfnisse und Forderungen des Marktes und der Kunden zu reagieren" (Schmelzer und Sesselmann, 2013, S 209). Kundenorientierung, Prozesseffizienz und Anpassungsfähigkeit sind auf ihrer Sicht die entscheidenden Erfolgsfaktoren.

Schmelzer und Sesselmann (2013, S. 210ff.) betrachten verschiedene Formen prozessorientierter Ausbauorganisationen mit ihren jeweiligen Vor-, aber auch Nachteilen. Sie propagieren aber auch den Ansatz „structure follows process follows strategy" (Schmelzer und Sesselmann, 2013, S. 220), wobei sie in der Praxis häufig die Umkehrung „process follows structure follows strategy" beobachten. Aus ihrer Sicht besteht hier aber entsprechend großer Handlungsbedarf in Unternehmen, um von dieser weit verbreiteten Praxis abzurücken.

Schmelzer und Sesselmann widmen in ihrem Buch zwar ein ganzes Kapitel der Thematik der IT-Unterstützung des Geschäftsprozessmanagements (Schmelzer und Sesselmann, 2013, S. 465ff.), diskutieren aber nicht die zukünftigen Einflüsse von Digitalisierung und Industrie 4.0 auf die Unternehmensorganisation. Da durch steigende Kundenanforderungen im Zuge kundenindividueller und personalisierter Produkte bzw. Losgröße 1 hier große Herausforderungen auf Unternehmen zukommen werden, bleibt abzuwarten, in wie weit sie ihre Ansätze weiter entwickeln werden.

Bis heute sind die meisten Unternehmen immer noch stark arbeitsteilig und hierarchisch organisiert. Diese Organisationsprinzipien sind aber in der neuen digitalen Welt überholt (Gassmann und Sutter, 2016). In der Zukunft werden sich also Unternehmens- und Produktionsstrukturen radikal ändern müssen, um den neuen Herausforderungen gewachsen zu sein. Heute sind die meisten Unternehmen immer noch klassisch funktional und hierarchisch organisiert mit einer tiefgreifenden Arbeitsteilung als Leitprinzip. Diese Prinzipien sind nach Gassmann und Sutter (2016, S. 10) jedoch im digitalen Zeitalter überholt. Sie schlagen 14 Maßnahmen vor, wie Unternehmen die digitale Transformation erfolgreich meistern können (Gassmann und Sutter, 2016, S. 11ff.). Vor allem die folgenden drei Maßnahmen stehen in einem engen Zusammenhang mit neuen Organisationsformen in Unternehmen:

- Da Digitalisierungsinitiativen fast immer funktions- und oft sogar unternehmensübergreifend sind, gilt es, bestehende funktionsorientierte Strukturen zu überwinden.
- Eine langfristige Planung sollte mehr und mehr durch Agilität, Lernen und Flexibilität ersetzt werden. Gerade in Zukunft wird es notwendig sein, rasch Neues auszuprobieren und bei Erfolglosigkeit wieder einzustellen. Dazu müssen schlagkräftige Projektteams mit Experten aus unterschiedlichen Disziplinen mit großen Entscheidungsfreiheiten geschaffen werden. Es gilt, „Zelte statt Paläste" auszubauen (Gassmann und Sutter, 2016, S. 13).
- Dies wird unterstützt durch die Förderung einer Lean-Start-up-Mentalität. Hier gilt es, Kundenfeedback zeitnah einzuholen und in die Weiterentwicklung von Produkten und Dienstleistungen einzubeziehen. Entwicklungs- und Projektphasen können so schneller durchlaufen werden.

3 Schlussbetrachtung und Fazit

Die Analyse des aktuellen state-of-the-art der wissenschaftlichen Diskussion hinsichtlich der Organisationsentwicklung zeigt die Bedeutung der Prozessorientierung in Unternehmen. Der Ansatz „structure follows process follows strategy" stellt auch nach Ansicht der Autoren den richtigen Anknüpfungspunkt für zukünftige Entwicklungen dar. Ausgehend von der Anfangsfrage, in wie weit die Digitalisierung und die Entwicklungen hin zu Industrie 4.0 Einfluss auf zukünftige Unternehmens- und Produktionsstrukturen nehmen werden, ist davon auszugehen, dass sämtliche Entwicklungen rund um die Digitalisierung und neue daraus resultierende Geschäftsmodelle die Fokussierung auf die Wünsche und Anforderungen der Kunden immer weiter vorantreiben werden. Da es aber gerade ein Vorteil prozessorientiert ausgerichteter

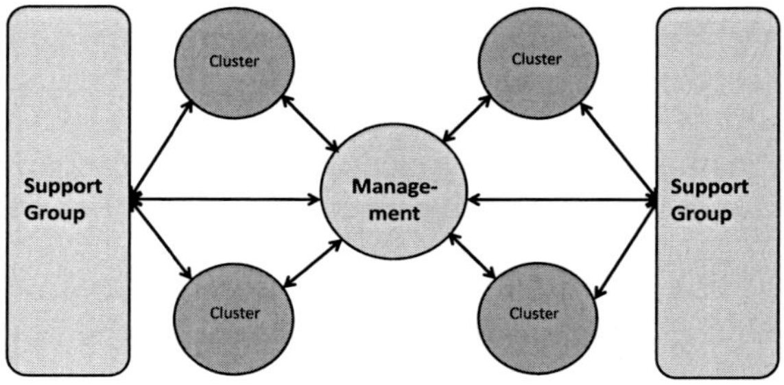

Abbildung 2: Clusterorganisation nach Gomez und Zimmermann (in Anlehnung an Gomez und Zimmermann (1993))

Unternehmen ist, in erster Linie die Kundenorientierung in den Mittelpunkt der Unternehmensführung zu stellen, wird die Entwicklung neuer Organisationsformen für Unternehmen eine große Zukunftsaufgabe werden. Ansätze, die hier ein vielversprechendes Potential zeigen, lassen sich mit den Schlagworten wie Netzwerke, virtuelle Organisationen, Allianzen, Schwarmintelligenz oder Clusterorganisationen beschreiben.

Staub (2015, S. 431f.) beschreibt eine Clusterorganisation als die Bildung autonomer Subsysteme, sogenannter Cluster, die weitgehend autonom und eigenverantwortlich agieren können. Sie werden von Support Groups unterstützt. Abbildung 2 veranschaulicht diese Organisationsform.

Da sie sehr stark auf Kollaboration und Kommunikation aufbaut, besitzt diese Organisationsform gerade im Zeitalter der Digitalisierung ein hohes Potential im Hinblick auf eine Stärkung der Wettbewerbsfähigkeit von Unternehmen mit Blick auf eine schnelle Reaktion auf sich ändernde Kundenanforderungen. Diesbezüglich können die Vorteile einer Clusterorganisation (Straub, 2015)

- Förderung des unternehmerischen Verhaltens der Mitarbeiter,
- hohe Flexibilität,
- gutes Klima für Innovation,
- variationsreicher, strategiekonformer Einsatz der Mitarbeiter sowie
- geringe Overhead-Kosten dank flacher Struktur

zum Tragen kommen.

Organisationsformen wie die Cluster Organisation können helfen, das Ziel einer prozessorientierten und gleichzeitig agilen Aufbauorganisation zu realisieren. Voraussetzung ist, dass es gelingt, die daraus resultierende Multidimensionalität der Aufgaben und Fähigkeiten von Mitarbeitern und Führungskräften zu beherrschen.

Es gilt heute Antworten zu finden, die die Nachteile hochgradig arbeitsteilig aufgestellter Unternehmensstrukturen nachhaltig überwinden. Damit werden zukünftig große Wettbewerbsvorteile zu erwarten sein.

Unterstrichen werden diese Schlussfolgerungen zum einen durch eine aktuell von den Autoren durchgeführte Befragung von ca. 40 Experten aus Unternehmen verschiedener Branchen und Größen. Auch sie sehen die Herausforderung, die Prozessorientierung in Unternehmen stärker auszugestalten und vor allem auch das herrschende Zuständigkeitsdenken zu überwinden.

Zum anderen zeigt ein Blick auf den typischen Prozess des Änderungsmanagements in vielen Unternehmen die Nachteile funktionsorientiert aufgestellter Unternehmen deutlich auf. Das klassische Änderungsmanagement mit dem sequentiellen Prozess beginnend mit der Marktbeobachtung über die Entwicklung, den Einkauf, die Produktion bis hin zu den unterstützenden Funktionen der IT und des Controllings ist zukünftig zu träge. Es ist aber zu erwarten, dass die meisten Unternehmen versuchen werden, sich evolutionär weiter zu entwickeln. Sollten aber Wettbewerber mit stark revolutionären, disruptiven Ansätzen auftreten, werden die Veränderungsprozesse möglicherweise nicht schnell genug sein, um sich weiterhin am Markt zu behaupten. Deswegen gilt es für Unternehmen, viel stärker proaktiv zu handeln und frühzeitig Ressourcen für ein umfassendes Change Management auszubauen.

Literatur

Bach, N.; Brehm, C.; Buchholz. W., Petry, T. (2017): Organisation - Gestaltung wertschöpfungsorientierter Architekturen, Prozesse und Strukturen. Springer Gabler, Wiesbaden

Bauernhansl, T. (2014): Die vierte Industrielle Revolution - Der Weg in ein wertschaffendes Produktionsparadigma. In: Bauernhansl, T.; ten Hompel, M.; Vogel-Heuser, B. (Hrsg.): Industrie 4.0 in Produktion, Automatisierung und Logistik, Springer Vieweg, Wiesbaden

Bieger, T.; Reinhold, S. (2011): Das wertbasierte Geschäftsmodell - Ein aktualisierter Strukturierungsansatz. In: Bieger T., zu Knyphausen-Aufseß, D., Krys, C. (Hrsg.): Innovative Geschäftsmodelle. Springer Verlag, Heidelberg.

Chandler, A. D. (1962): Strategy & Structure. MIT Press, Cambridge, USA

Gassmann, O.; Sutter, P. (2016): Software erobert die Welt. In: Gassmann, O., Sutter, P. (Hrsg.): Digitale Transformation in Unternehmen gestalten. Carl Hanser Verlag, München.

Gomez, P., Zimmermann, T. (1993): Unternehmensorganisation: Profile, Dynamik, Methodik. 2. Auflage, Campus Verlag, Frankfurt.

Koren, Y. (2010): The Global Manufacturing Revolution: Product-Process-Business Integration and Reconfigurable Systems. John Wiley & Sons, New York, USA.

Kosiol, E. (1962): Organisation der Unternehmung. Gabler Verlag, Wiesbaden

Lichtblau, K.; Stich, V.; Bertenbach, R.; Blum, M.; Bleider, M.; Millack, A.; Schmitt, K.; Schmitz, E.; Schröter, M. (2015): Industrie 4.0-Readiness. Institut der deutschen Wirtschaft, Köln und FIR e.V. an der RWTH Aachen.

Schmelzer, H. J.; Sesselmann, W. (2013): Geschäftsprozessmanagement in der Praxis. 8. Auflage, Carl Hanser Verlag, München.

Straub, T. (2015): Einführung in die Allgemeine Betriebswirtschaftslehre. 2. Auflage, Pearson Deutschland, Halbergmoos.

Shopfloor Management in der selbstgesteuerten Logistik-Lern-Fabrik

Stefan Faller[a], Marcus Hampa[a], Günter Bitsch[b], Vera Hummel[a], Beate Brenner[a]

[a] *ESB Business School, Reutlingen University, D-72762 Reutlingen*
[b] *becos GmbH, Zetachring 2, 70567 Stuttgart, Germany*

Abstract

Gesellschaftliche und industrielle Trends im Zuge der Digitalisierung induzieren Veränderungsprozesse in der Industrie. Eine hohe Flexibilität und schnelle Entscheidungsfindungsprozesse stellen entscheidende Wettbewerbsvorteile für Unternehmen dar, um zukünftig erfolgreich am Markt agieren zu können. Um dies zu ermöglichen, müssen aggregierte Echtzeitdaten und Prognosen unmittelbar sowohl am Ort der Wertschöpfung, als auch dezentral zur Verfügung stehen. Die Entscheidungsunterstützung mit Hilfe geeigneter Visualisierungen ist ein maßgeblicher Bestandteil von Shopfloor Management Systemen. Aufgrund der steigenden Anforderungen wurde das konventionelle und analoge Shopfloor Management in den letzten Jahren verstärkt durch digitale Lösungen ersetzt. Ein ganzheitlicher Shopfloor Management Ansatz, der die Trends und die daraus resultierenden Herausforderungen für die Industrie abdeckt, ist aktuell nicht vorhanden. Zukünftige Shopfloor Management Lösungen sollen diese Lücke schließen. Hierfür wurde ein ganzheitliches System entwickelt, welches Produktionsinformationen in Echtzeit unmittelbar am Shopfloor visualisiert, eine integrierte flexible Planung und Steuerung der Produktion beinhaltet sowie die Mitarbeiterbedürfnisse berücksichtigt. Eine flexible und individuelle Schichtplanung durch die Mitarbeiter und eine umfassende automatische Beanspruchungsbeurteilung sind dazu integriert worden. Zudem ermöglicht das System die Prognose und Visualisierung von Produktionsinformationen und unterstützt die Anwender bei der Durchführung strukturierter Shopfloor-Meetings. Dadurch werden Entscheidungen direkt auf den Ort der Wertschöpfung verlagert.

1 Einleitung

Die globalen Märkte und Produktionsumgebungen werden in Zukunft von signifikanten industriellen und gesellschaftlichen Trends beeinflusst (Spath et al 2013). „Industrie 4.0", die „Generation Y" und die anhaltende Globalisierung werden in den kommenden Jahren große Veränderungen in der Industrie nach sich ziehen.

Einer der größten sozialen Einflüsse für Unternehmen war der Eintritt der so genannten „Generation Y" in den Arbeitsmarkt. Diese Generation, welche zwischen 1980 und 2000 geboren ist, strebt nach einer ausgeglichenen Work-Life-Balance, welche flexible Arbeitsbedingungen voraussetzt (Parment 2009). Um das volle Potenzial dieser Mitarbeiter zu nutzen und um weiterhin erfolgreich am Arbeitsmarkt agieren zu können, ist es notwendig, neue flexible Rahmenbedingungen für diese Mitarbeiter zu schaffen (Diabaté und Bujard, 2015).

Flexibilität ist aber auch die Grundvoraussetzung, um an den immer volatileren globalisierten Märkten, die durch steigende Kundenanforderungen und internationale Vernetzung verursacht werden, weiterhin bestehen zu können (Spath et al 2013). Der Begriff „Industrie 4.0" beschreibt die vierte industrielle Revolution, die sich vor allem damit beschäftigt, modernste Informations- und Kommunikationstechnologien mit Produktions- und Logistiktechnologien zu verbinden, um diese noch effizienter zu gestalten und den wachsenden Herausforderungen der Märkte gerecht zu werden (o.V., 2017). Kürzere Produktlebenszyklen, sich ständig ändernde Kundenanforderungen und komplexere Produkte sind nur einige dieser Herausforderungen, die es notwendig machen, schnellere und bessere Entscheidungen zu treffen (Westkämper, 2009). Eine Studie des Fraunhofer IAO, die sowohl Unternehmensvertreter als auch Industrie 4.0 Experten befragte, ergab das zusammengefasste „Bild der zukünftigen Produktionsarbeit" in Abbildung 1.

Themen, wie beispielsweise die Verarbeitung und Darstellung von Echtzeitdaten, wachsende Mobilität, voranschreitende Vernetzung und Dezentralität werden laut dieser Studie die Produktion der nächsten Jahre prägen. Trotz dieser Entwicklungen bestätigt die Studie aber auch, dass der Mensch in der Fabrik präsent bleibt und seine Bedürfnisse immer mehr in den Mittelpunkt rücken (Spath et al 2013). Vor dem Hintergrund der zusätzlich aufgeführten technologischen Trends (vgl. Abbildung 1) ist davon auszugehen, dass sich die Tätigkeiten des Menschen in der Fabrik verändern werden. Eine der wichtigsten Aufgaben der Mitarbeiter wird es sein, die Umsetzung einer vorgegebenen Produktionsstrategie innerhalb von selbst organisierten Produktionsprozessen zu überwachen (Gorecky et al, 2017). Um diese Überwachung zu ermöglichen, müssen Echtzeitdaten der Produktion am Shopfloor gesammelt, aggregiert und analysiert werden (Gröger et al, 2016). Mit diesen Daten kann eine bessere und aktuellere Entscheidungsgrundlage für alle Akteure geschaffen werden. Hierdurch wird ermöglicht, Entscheidungen schneller und in einer besseren Qualität zu treffen (Spath et al 2013). Dabei wird die umfassende Vernetzung und die Möglichkeit, mobil auf Informationen zuzugreifen, dazu führen, dass der ortsgebundene Arbeitsplatz zunehmend an Bedeutung verlieren wird und die Entscheidungs- und Überwachungsprozesse sowohl unmittelbar am Shopfloor, als auch dezentral und mobil erfolgen können (Gorecky et al, 2017).

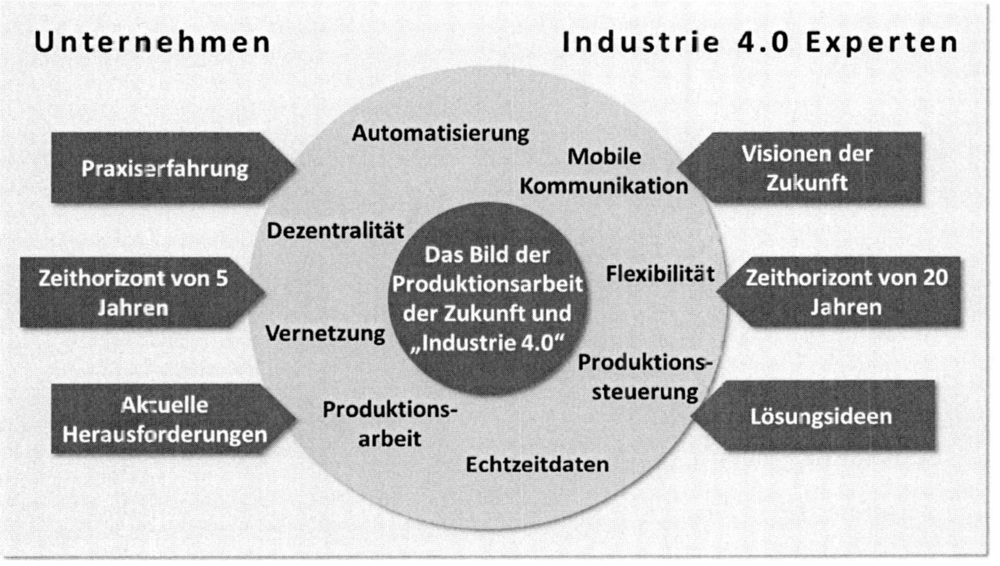

Abbildung 1: Bild der zukünftigen Produktionsarbeit. Eigene Darstellung in Anlehnung an Spath, D. (2013)

2 Stand der Technik

2.1 Konventionelles Shopfloor Management

Die Überwachung der Produktion und die damit einhergehende Entscheidungsunterstützung ist typischerweise die Aufgabe des Shopfloor Managements (SFM). Eine einheitliche Definition für das SFM existiert nicht. Vielmehr lässt es sich über die in Abbildung 2 dargestellten nachfolgenden Merkmale und Ziele definieren.

Beim SFM handelt es sich um ein visuelles Entscheidungsinstrument, welches den Mitarbeiter zur Optimierung von Produktionsprozessen am Ort der Wertschöpfung – dem Shopfloor – einbindet. In der Regel werden definierte Kennzahlen auf einer Tafel abgebildet, welche Probleme bzw. Potentiale aufzeigen und als Entscheidungsgrundlagen dienen. Operative und führende Mitarbeiter besprechen gemeinsam die Mitarbeiter- und Maschinenperformance und ermöglichen hierdurch eine effiziente Problemlösung und eine zielführende Erschließung des möglichen Potentials. Beim konventionellen Shopfloor Management wird in der Regel eine Magnettafel oder eine Pinnwand zur Visualisierung verwendet. Dort werden zu jedem Meeting die Kennzahlen des Vortags aus den Abteilungen mitgebracht, angeheftet und besprochen (Lendzian und Martin-Martin, 2016). Die Erstellung der Kennzahlen ist oft mit hohem manuellem Aufwand verbunden.

	Anstadt (2009)	Suzaki (1993)	Flores (o.J.)	Peters (2009)	Hurtz; Stolz (2013)	Hertle et al. (2015)
Schaffung von Transparenz	X	X	X		X	X
Strukturierte Vorgehensweise	X		X			
Abweichungen erkennen und Probleme lösen		X	X	X		X
Führen vor Ort	X	X	X	X	X	X
Prozessoptimierung		X			X	X
Mitarbeiter einbeziehen		X	X	X	X	X

Abbildung 2: Merkmale und Ziele SFM. Eigene Darstellung in Anlehnung an Anstadt, U (2009). Suzaki, K. (1993). Flores, K. (o.J.). Peters, R. (2009). Hurtz, A.; Stolz, M. (2013). Hertle, C. et al. (2015).

2.2 Digitales Shopfloor Management

Im Gegensatz zum konventionellen SFM wird beim weiterentwickelten digitalen SFM ein Bildschirm als Visualisierungstool herangezogen. Für diese Art der Darstellung gibt es bereits einige im Folgenden zusammengefasste Konzepte. Die Lösungen für ein digitales Dashboard unterscheiden sich hauptsächlich in der adressierten Hierarchieebene (Gröger et al, 2016). Das so genannte „Business Activity Monitoring" (McCoy, 2002) beschäftigt sich mit der echtzeitbasierten Datenerfassung und Auswertung zur Erkennung von Prozessabweichungen. Anwendung finden diese Systeme hauptsächlich auf der obersten Hierarchieebene zur Überwachung von Kennzahlen (Muehlen und Shapiro, 2010). Im Gegensatz dazu werden die so genannten „MES Cockpits" hauptsächlich von Produktionsmanagern in der Produktionssteuerung verwendet. Sie sind üblicherweise ein Teil des Manufacturing Execution System (MES) und unterstützen den Produktionsmanager bei den Themen Produktionsplanung, Ressourcenplanung und Prozessüberwachung (Kletti, 2007). Speziell für den Shopfloor wurde von Bracht, Hackenberg und Bierwirth ein „Visualisierungssystem für operative Logistik" entwickelt (Bracht et al, 2011). Es wurde als Applikation auf einem Tablet-PC für Logistik-Mitarbeiter implementiert und zeigt hauptsächlich echtzeitbasierte Kennzahlen an, welche einen Überblick über die Prozessperformance liefern sollen (Gröger et al, 2016).

Als Erweiterung für diese – auf wenige Teilaspekte spezialisierten – Konzepte haben Gröger et al. einen ganzheitlichen Ansatz für ein mobiles Dashboard entwickelt. Das „Operational Process Dashboard for Manufacturing" (OPDM) stellt nicht nur simple kennzahlenbasierte Informationen dar, sondern analysiert mit Hilfe von Data-Mining-Diensten die gesamten prozessorientierten Informationen (Gröger et al, 2016). Ein Vorteil dieser digitalen Lösungen liegt darin, dass die Generierung der Kennzahlen vollautomatisch und auf Basis von Echtzeitdaten erfolgt. Weiterhin wird dadurch die mobile und dezentrale Abrufbarkeit der Informationen ermöglicht.

3 Anforderungen an zukünftige Shopfloor Management Lösungen

3.1 Forschungsbedarf

Alle dargestellten digitalen Konzepte konzentrieren sich auf das Schaffen von Transparenz, das Erkennen von Abweichungen und die daraus ableitbare Prozessoptimierung. Damit stellen sie keinen ganzheitlichen Shopfloor Management Ansatz dar, da sie sowohl den strukturierten Ablauf eines solchen Meetings, als auch das Führen am Ort der Wertschöpfung nicht miteinbeziehen. Auch das elementare Berücksichtigen der Ideen und der Kreativität der auf dem Shopfloor arbeitenden operativen Mitarbeiter bleibt unbeachtet.

Ein weiterer Nachteil besteht darin, dass die Mitarbeiter bei den digitalen Lösungen nicht mehr aktiv an der Erstellung der Informationen beteiligt werden und damit eine Entfremdung von den angezeigten Inhalten stattfinden kann. Dies ist einer der Hauptgründe, warum im SFM trotz fortschreitender Digitalisierung immer noch auf die konventionelle Variante mit Magnettafeln und Pinnwänden vertraut wird (Schöllhamer, 2013). Des Weiteren werden die in Abbildung 1 aufgezeigten industriellen Entwicklungen unter dem Einfluss von Industrie 4.0 nur teilweise berücksichtigt. Die erhöhten Anforderungen an die Flexibilität, die in der Einleitung erläutert wurden, werden nicht miteinbezogen. Die steigenden Anforderungen sowohl der Kunden, als auch der Mitarbeiter, erfordern laut einer Studie des Fraunhofer IAO den Einsatz neuer Instrumente der Kapazitätsplanung (Spath et al 2013).

Die Verarbeitung und Visualisierung von Vergangenheits- und Echtzeitdaten wird in den Konzepten der digitalen Dashboards bereits berücksichtigt. Prognosetechniken und Simulationen zur Entscheidungsunterstützung werden selten angewendet.

Ein weiterer Aspekt, der nicht im Fokus der aktuellen Lösungen steht, ist die Einbeziehung des Menschen als zentralen Faktor des SFM. Die Mitarbeiter sind durch die steigenden Anforderungen einer immer größeren Belastung ausgesetzt. Dabei sollen gerade diese in Zukunft wichtige Funktionen in der digitalen Fabrik einnehmen. Daher sollte auch die Betrachtung der Arbeitsbedingungen in den Fokus des Shopfloor Managements rücken.

Abbildung 3: Ziele des zukunftsorientierten digitalen SFM

3.2 Ziele

Abgeleitet von den aktuellen Trends, dem „Bild der Produktionsarbeit von morgen" und den aufgeführten Nachteilen der aktuellen digitalen Lösungen ist es das Ziel, ein neues ganzheitliches SFM-Konzept zu entwickeln, das alle genannten Aspekte miteinbezieht. In Abbildung 3 sind die Zielsetzungen eines zukunftsorientierten digitalen SFM-Konzepts dargestellt.

Das Fundament für das entwickelte Konzept bilden die bisherigen Ansätze des digitalen Shopfloor Managements. Im Fokus liegt neben der Erfüllung von Kundenwünschen die Berücksichtigung der Bedürfnisse der Mitarbeiter. Zentral werden die Ziele und Visionen des in Kapitel vier erläuterten ganzheitlichen Ansatzes des digitalen SFM aufgeführt. Dazu gehören die Betrachtung zukunftsbezogener Daten, die Vorbeugung vor Fehlbeanspruchung, die Simulation des Kapazitätsbedarfs, die Visualisierung aller Prozesse der digitalen Fabrik unter Gewährleistung einer flexiblen Personal- und Auftragsplanung. Zusätzlich sind sowohl ein Ideenmanagement zur Nutzung der Mitarbeiterkreativität, als auch ein Fehlermanagement zur wirksamen sowie nachhaltigen Problemlösung entwickelt und integriert. Ein zusätzlicher Aspekt ist auch die Berücksichtigung der Beanspruchung der Mitarbeiter. All diese Funktionen werden auf einem digitalen Dashboard direkt auf dem Shopfloor visualisiert und ermöglichen eine Interaktion.

4 Umsetzungskonzept

4.1 Forschungsumgebung

Ausgehend von den genannten gesellschaftlichen und industriellen Trends galt es, ein Konzept für zukünftige Shopfloor Management Lösungen zu entwickeln. Dies erfolgte im Rahmen eines wissenschaftlichen Projekts im Wintersemester 2016/17 innerhalb des Masterstudiengangs „Operations Management (M.Sc.)" der ESB Business School Reutlingen in Zusammenarbeit mit der Firma becos in Stuttgart. Als Forschungs- und Validierungseinrichtung wurde die Umgebung der Logistik-Lern-Fabrik (LLF) der ESB Business School genutzt. Innerhalb der LLF befindet sich eine Produktionsumgebung für die Fertigung variantenreicher Produkte durch innovative Technologien (Schumacher und Hummel, 2016). Lernfabriken bieten dabei ideale Voraussetzungen für interdisziplinäre Forschung zu zukünftig relevanten Themenstellungen. Seit der Einführung von „Lern-Fabriken" im Jahr 1994 bieten diese eine Lern-, Ausbildungs- und Forschungsumgebung für Studenten und Berufserfahrene, die auf realen industriellen Produktions- und Logistiktechnologien basieren (Abele et al, 2015).

4.2 „Digital Interactive Dashboard" und „Mitarbeiter Cockpit"

Ähnlich der Struktur des OPDM basiert das System auf der Trennung von Visualisierung im „Frontend" und der Datensammlung beziehungsweise -analyse im „Backend" (Gröger et al, 2016). Für die Visualisierung und Interaktion am Shopfloor wird als Frontend-Lösungen ein mobiles Microsoft Surface Hub (MSH) mit 84" Touch-Screen als „Digital Interactive Dashboard" (DID) eingesetzt. Zusätzlich werden unterschiedliche mobile Endgeräte auf Windows- und Android-Basis eingesetzt. Im Backend erfolgt die erforderliche Datenanalyse innerhalb des Cloud-Dienstes Microsoft Azure. Das DID, welches unmittelbar am Shopfloor eingesetzt wird, zeigt auf dem Startbildschirm dynamisch die relevanten Kennzahlen der Produktion sowie die Auftragsplanung für einen definierten Zeitraum inklusive wichtiger Auftragsinformationen. Zusätzlich werden aktuelle Informationen für alle Mitarbeiter dargestellt. Über den „Home-Button" können alle zusätzlich entwickelten Anwendungen, beispielsweise für die detailliertere Darstellung der KPIs oder zum Abruf des Schichtplans, gestartet werden.

Alle Funktionen können auch dezentral über Windows-basierte Systeme aufgerufen und genutzt werden. Damit wird das Ziel der dezentralen Verfügbarkeit des Systems erfüllt. Den Mitarbeiter steht als zentrales Interaktionsmedium die Smartphone-Applikation „Mitarbeiter Cockpit" (MiCo) zur Verfügung, die auf private Smartphones mit Android-Betriebssystem heruntergeladen werden kann. Dies basiert auf dem Konzept „Bring your own Device". Die App beinhaltet die vier Reiter „Beanspruchung", „Schichtplanung", „Ideenmanagement" und „Benachrichtigungen".

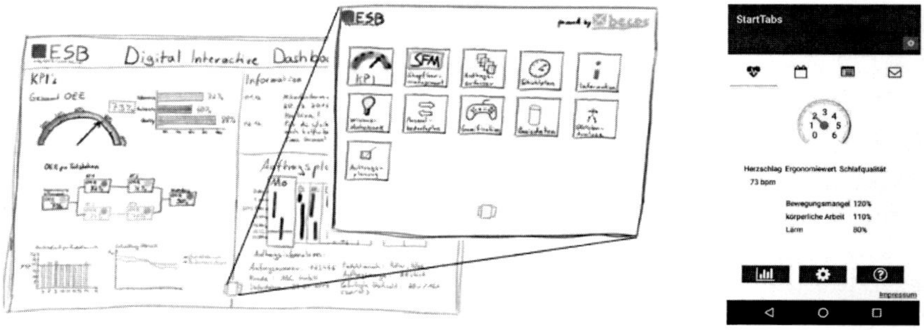

Abbildung 4: Digitales interaktives Dashboard (links) und Mitarbeiter Cockpit (rechts)

4.3 Beanspruchung

Ein Ziel des SFM-Konzepts ist es, aufgrund der immer dynamischeren Anforderungen an die Mitarbeiter die individuelle Beanspruchung zu berücksichtigen. Die Unternehmen sind angehalten, proaktiv und vorbeugend Maßnahmen einzuleiten, um ihre Belegschaft vor Fehlbeanspruchung zu schützen. Um dieser Aufgabe gerecht zu werden, bedarf es einer ganzheitlichen Beanspruchungsbeurteilung mit Berücksichtigung aller Belastungsfaktoren. Die Beanspruchung eines Mitarbeiters ist die „individuelle Auswirkung der Arbeitsbelastung des Menschen in Abhängigkeit seiner Eigenschaften und Fähigkeiten" (Hummel, o.J.). Um die Beanspruchung zu erfassen, ist es notwendig, die Arbeitsbelastung und die individuelle Leistungsfähigkeit des Menschen zu messen. Die beiden Faktoren müssen dabei in Relation zueinander gesetzt werden, um daraus die Beanspruchung ableiten zu können. Die Leistungsfähigkeit ist nach DIN33400 als „der Faktor, mit dem die Beanspruchung mit der Belastung verknüpft ist" definiert. Die Arbeitsbelastung ist definiert als „die Gesamtheit der erfassten Einflüsse, die im Arbeitssystem auf den Menschen einwirken" (Hummel, o.J.). Entscheidend sind dabei die Dauer und die Höhe der Belastung (Hummel, o.J.). Um eine Analyse zu ermöglichen, geben die Mitarbeiter im Reiter „Beanspruchung" ihre persönlichen Daten an. Neben den Basisinformationen, wie beispielsweise dem Alter oder dem Gewicht des Nutzers, wird hierbei auch die grundsätzliche Sensibilität gegenüber Umwelteinflüssen, wie gegenüber Temperaturänderungen, hinterlegt. Diese Informationen können jederzeit bedarfsgerecht angepasst werden. Um variable Faktoren zu berücksichtigen, dient ein Fragebogen, der die Tagesform, die Laune, die Motivation, die Befindlichkeit und den Stress des Nutzers abfragt. Dieser wird von den Mitarbeitern täglich vor Arbeitsbeginn über das „Mitarbeiter Cockpit" ausgefüllt. Die daraus gewonnenen Daten geben Aufschluss über die individuelle psychische Beanspruchung der Mitarbeiter. Neben der psychischen Beanspruchung, wird die physische Beanspruchung der Mitarbeiter über Sensoren ermittelt. Zum einen wird ein Fitness-Tracker eingesetzt, der die relevanten Daten des Trägers, wie den Puls oder Werte über die Schlafqualität, ermittelt. Zum anderen wird zur Beurteilung der Ergonomie eine Microsoft Kinect für die Analyse von Bewegungen und der Körperhaltung während der Produktion eingesetzt. Das System arbeitet mit Infrarotsensoren und liefert Raumkoordinaten von 26 unterschiedlichen Ortungspunkten des menschlichen Körpers. Hierdurch kann die Ergonomie am Arbeitsplatz ermittelt und damit in die individuelle Analyse der Beanspruchung einbezogen werden. Nicht ergonomische Haltungen und Bewegungen, wie das Heben über Schulterhöhe oder das Verdrehen der Wirbelsäule, werden vom System durch die stetige Aufnahme erkannt und beeinflussen die dynamische Beanspruchung über den Arbeitstag.

Weitere relevante Faktoren ergeben sich aus den Umwelteinflüssen innerhalb der Produktionsumgebung. Hierfür werden weitere Sensoren eingesetzt, die diese Einflüsse an jedem Arbeitsplatz stetig überwachen. Gemessen werden die Temperatur, die Helligkeit, die Luftfeuchtigkeit, der Luftzug, der CO2-Gehalt der Luft sowie Schwingungen und Vibrationen.

Alle Beanspruchungen werden zu einer Gesamtbeanspruchung aggregiert, die auch zur Prognose von Kennzahlen herangezogen werden. Zusätzlich können die jeweiligen Einzelbeanspruchungen jederzeit über das MiCo von den Mitarbeitern abgerufen werden. Bei Überschreiten definierter Schwellwerte werden die betroffenen Mitarbeiter per Push-Benachrichtigung auf die Überbelastung hingewiesen.

Die Verarbeitung der gewonnenen Daten dient maßgeblich dem Schutz und der Unterstützung der Mitarbeiter und stellt keine Grundlage für Überwachungstätigkeiten dar. Persönliche Informationen und die individuellen Einzelbeanspruchungen können daher nur vom Mitarbeiter selbst abgerufen werden. Auf dem „Digital Interactive Dashboard“ stehen nur aggregierte Gesamtinformationen zur Verfügung, die keine direkten Rückschlüsse auf einzelne Mitarbeiter erlauben.

4.4 Flexible mitarbeiterorientierte Planung

Ein weiteres Ziel ist die flexible Personal- und Auftragsplanung, welche eine situationsbedingte Änderung am Shopfloor ermöglicht. Zur Ermittlung des Mitarbeiterbedarfs und zur Optimierung des Ressourcenmanagements beinhaltet das System ein Prognosewerkzeug, welches anhand durchgeführter Simulationen die optimale Mitarbeiterkapazität ermittelt.

Um die größtmögliche Flexibilität für die Produktionsmitarbeiter zu erreichen, werden diese nicht mit Hilfe starrer Planungstools eingeplant, sondern legen ihre jeweiligen Schichten selbständig fest. Die Festlegung der Schichten erfolgt über den Reiter „Schichtplanung“ des „Mitarbeiter Cockpits“, wobei dem Nutzer hierdurch der Zugriff auf seinen individuellen Schichtplan ermöglicht wird. Hier werden die vom Mitarbeiter gewünschten Arbeitszeiten für einen definierten Zeitraum eingetragen und grafisch dargestellt. Zusätzlich wird hier die gewünschte Gesamtarbeitszeit für den Planungszeitraum festgelegt. Der Planungszeitraum kann bedarfsgerecht angepasst werden. Dieser sollte so festgelegt werden, dass ein Kompromiss zwischen Flexibilität der Produktion und Planungssicherheit der Mitarbeiter erreicht werden kann. Lange Planungshorizonte führen durch die Berücksichtigung von Auswirkungen auf zukünftige Perioden zu einer höheren Planungsgenauigkeit. Durch kürzere

Planungsabschnitte können jedoch Änderungen aufgrund interner oder externer Ereignisse innerhalb der Fertigung berücksichtigt werden (Steven, 2013).

Um die Einzelschichtpläne der Mitarbeiter aus dem MiCo zu aggregieren und daraus einen Gesamt-Schichtplan für den im Vorfeld definierten Planungszeitraum zu erstellen, erfolgt eine lineare Optimierung aller eingetragenen Mitarbeiterschichten. Der Zielwert für die Berechnung ist die höchste Übereinstimmung der von allen Mitarbeitern festgelegten Arbeitszeiten unter Berücksichtigung der individuell festgelegten Wunscharbeitszeit. Bei der Erstellung der Schichtplanung wird zudem die erwartete Bearbeitungszeit der Aufträge berücksichtigt. Dadurch wird sichergestellt, dass die Mitarbeiter nur für vorhandene Aufträge eingeplant werden. Auf Basis des Gesamt-Schichtplans werden die Kundenaufträge nach vordefinierten Priorisierungsregeln auf die Schichten verteilt und der Auftragsplan vom System automatisch generiert. Bei Bedarf können zusätzlich manuelle Änderungen, wie die Anpassung der Auftragsdaten, die Änderung der Priorisierung oder die Einplanung zusätzlicher Aufträge, durchgeführt werden. Nach Abschluss der Auftragsplanung wird der generierte Plan mit allen relevanten Mitarbeiter- und Auftragsinformationen am DID dargestellt.

Neben der einfachen Darstellung wird über eine Applikation des „Digital Interactive Dashboard“ jederzeit ermöglicht, Änderungen unmittelbar am Shopfloor durchzuführen. Steuernde Eingriffe auf die Planung und die damit einhergehenden Auswirkungen auf den Verlauf der Produktion sind zudem unmittelbar am DID ersichtlich und dienen damit als Entscheidungshilfe für den jeweiligen Anwender. Externe und unvermeidbare Einflüsse, wie beispielsweise Störungen oder Mitarbeiterausfälle, werden vom System automatisch berücksichtigt. Dabei wird die Planung dynamisch angepasst und alle betroffenen Mitarbeiter werden automatisch benachrichtigt.

Um die Flexibilität der Mitarbeiter innerhalb des definierten Planungszeitraums weiter zu erhöhen, können die festgelegten Schichten bei Bedarf abgegeben werden. Dies erfolgt über den Reiter „Benachrichtigung“ des MiCo. Hier können die geplanten Schichten ausgewählt und eine individuelle Anfrage gestartet werden. Alle qualifizierten Mitarbeiter, die im ausgewählten Zeitraum nicht bereits eingeplant sind, können dieser Anfrage zustimmen. Die daraus resultierenden Personaländerungen werden direkt in der Auftragsplanung am DID berücksichtigt.

4.5 Dynamische Visualisierung von Kennzahlen

Eine zentrale Funktion des „Digital Interactive Dashboards“ ist es, Produktionsinformationen situationsbedingt unmittelbar am Shopfloor zu visualisieren. Die Darstellung passt sich dabei variabel dem individuellen Informationsbedarf des aktuellen Nutzers an Der anwenderspezifische Zugriff erfolgt über den Einsatz von Gesichtserkennungssoftware. Die Visualisierung aller Kennzahlen basiert auf Vergangenheits- und Echtzeitdaten, die automatisiert aus der vernetzten Produktion gewonnen werden. Aus diesen Daten werden Prognosen abgeleitet, um bedarfsgerecht zukunftsorientierte Entscheidungen am Shopfloor treffen zu können.

Die zentrale Kennzahl auf dem DID ist die Gesamtanlageneffektivität (OEE), welche sich aus der Performance, der Verfügbarkeit und der Qualität zusammensetzt. Die

OEE wird für die gesamte Produktion sowie für die einzelnen Arbeitsplätze ermittelt, visualisiert und prognostiziert. Weitere relevante Kennzahlen für die Produktion, wie der Durchsatz, die Durchlaufzeit oder die Häufigkeit der Störungen, werden ebenfalls am DID in Echtzeit dargestellt und prognostiziert.

Neben den technischen Aspekten wird zudem das Qualifizierungslevel der jeweiligen Mitarbeiter fortlaufend ermittelt. Aufgrund der gewonnen Mitarbeiterinformationen kann die Bearbeitungszeit der zu fertigenden Produkte in Abhängigkeit vom Qualifizierungslevel und der individuellen Beanspruchung des Mitarbeiters prognostiziert werden.

Die Arbeitsplätze der Mitarbeiter sind mit Tablet-PCs ausgerüstet, die als Informations- und Interaktionstool dienen. Darauf werden, abhängig von der aktuellen Produktvariante, die per RFID-Technologie automatisch erkannt wird, Utility-Filme als Montageanweisung dargestellt. Darüber hinaus ist ein Fehler- und Störungsmanagementsystem implementiert. Fehler und Störungen werden nach Produkt und Produktionsressourcen unterschieden. Auftretende Produktfehler können von den Mitarbeitern anhand des eingebundenes 3D CAD Modells spezifiziert werden und digital mit allen benötigen Informationen an das Lagerverwaltungssystem gemeldet werden. Die am jeweiligen Arbeitsplatz verfügbaren Produktionsressourcen sind durch Schnittstellen mit dem System verknüpft. Bei Störungen können die betroffenen Fertigungsressourcen aus einer Übersicht der jeweils vorhandenen Ressourcen ausgewählt werden. Unmittelbar im Anschluss erfolgt eine automatische Benachrichtigung des zuständigen Instandsetzungsmitarbeiters. Neben den hinterlegten Informationen können zudem zusätzliche Details per Videoaufnahme oder Sprachnachrichten des Mitarbeiters übermittelt werden.

4.6 Shopfloor Management Meeting

Das „Digital Interactive Dashboard" ist zusätzlich das zentrale Werkzeug zur Durchführung von SFM-Meetings. Für die Unterstützung eines strukturierten Ablaufs ist hierfür eine Anwendung im System integriert, die eine Agenda mit fünf Schritten beinhaltet.

Im ersten Schritt werden die aktuellen und produktionsrelevanten Kennzahlen situationsbedingt dargestellt, wobei kritische Werte hervorgehoben werden. Der Zeitraum der Betrachtung kann vom Anwender individuell festgelegt werden und reicht von wenigen Stunden bis zu mehreren Tagen. Eine Änderung der Festlegungen wirkt sich dynamisch auf die grafische Darstellung aus. Neben den initial dargestellten Kennzahlen kann jeder Einzelwert im Detail betrachtet und analysiert werden, um dem Anwender die Ursachenanalyse zu vereinfachen. Bei aufgetretenen Problemen können Verbesserungsideen oder Lösungsvorschläge durch Sprach- oder Texteingaben aufgenommen werden und unmittelbar die Zuständigkeiten mit Hilfe des Organigramms festgelegt werden.

Im zweiten Schritt wird die Auftragsplanung über einen vordefinierten Zeitraum dargestellt. Hierbei können manuell Änderungen, wie beispielsweise die Einplanung

Abbildung 5: Mobiles interaktives Board, operatives Mitarbeiter-Shopfloor-Meeting links und Management-Shopfloor-Meeting rechts

zusätzlicher Aufträge oder die Änderung von Mitarbeiterzuteilungen aufgrund hoher Beanspruchungswerte, durchgeführt werden. Kritische Aufträge, die nicht zur geplanten Lieferzeit fertig gestellt werden, sind in dieser Darstellung der Auftragsplanung hervorgehoben. Alle durchgeführten Änderungen sind reversibel und beeinflussen die Schichtplanung erst durch die Freigabe der Eingaben.

Im dritten Schritt werden alle neuen, umgesetzten und für wirksam erklärten Mitarbeiterideen dargestellt und jeweils besprochen. Das Ziel ist die Sicherstellung der nachhaltigen und wirksamen Umsetzung der Ideen. Dazu können, unabhängig von der jeweiligen Organisationseinheit, interdisziplinäre Teams zur Umsetzung der Ideen gebildet werden.

Durch das im „Mitarbeiter Cockpit" integrierte „Ideenmanagement" wird den Mitarbeitern ermöglicht, Verbesserungsvorschläge schnell, einfach und strukturiert abzugeben. Die Anlage neuer Ideen erfolgt anhand von Beschreibungen, die auch per Spracherkennung generiert werden können. Zusätzlich können den Beschreibungen Bilder und Videos zur Veranschaulichung hinzugefügt werden. Zudem kann sowohl das mögliche Potential, als auch das Investitionsvolumen abgeschätzt und eingetragen werden. Dies ermöglicht eine umfassende und gemeinschaftliche Beurteilung der Ideen während des SFM-Meetings. Außerdem ist allen Mitarbeitern die Möglichkeit gegeben, über diesen Reiter alle Ideen mit dem jeweiligen Status und Bearbeitungsstand einzusehen.

Im vierten Schritt, dem Fehlermanagement, werden die durch Fehler und Probleme induzierten Verbesserungsprozesse aggregiert dargestellt. Das Ziel ist, kollektiv eine Lösung für aufgetretene Probleme zu finden. Zudem erfolgen hierbei die Überprüfung der Wirksamkeit und die Kontrolle der Nachhaltigkeit der durchgeführten Maßnahmen. Der Ablauf dieses Schritts orientiert sich am strukturierten und teamorientierten Lösungsprozess eines 8D-Reports.

Im fünften und letzten Schritt werden alle vorangegangen Schritte in einer Übersicht dargestellt. Zudem ist der Projektstatus des Ideen- und Fehlermanagements für die Teilnehmer ersichtlich. Durch das Beenden der Anwendung wird diese Übersichtsdarstellung automatisch als Protokoll im System hinterlegt und an alle relevanten Zielgruppen weitergeleitet.

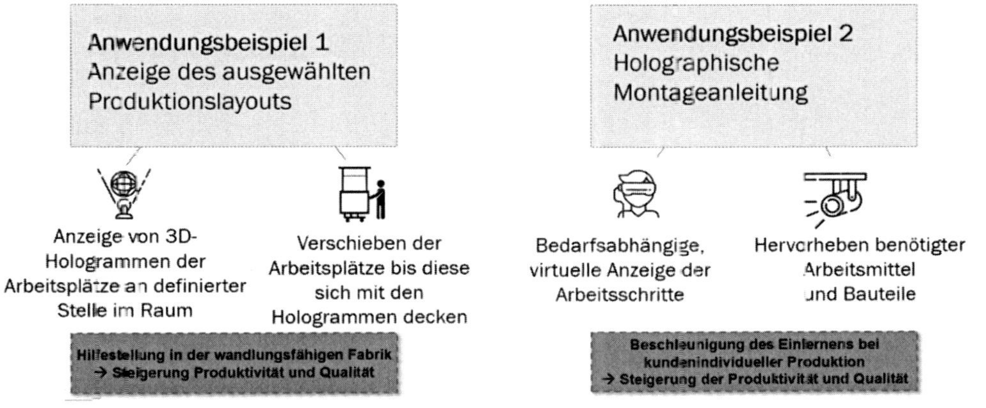

Abbildung 6: Anwendungsbeispiele der automatisierten Informationsbereitstellung durch AR-Technologie

4.7 Automatisierte Informationsbereitsstellung durch Augmented-Reality AR - Technologie

Zur automatisierten Informationsbereitstellung für wandlungsfähige Produktionslayouts je nach Fertigungsstückzahl und Ressourcenverfügbarkeit werden direkt aus der Businessplattform (enthält alle Daten der Fabrik online in der Cloud, App-basierte Anwendungen) Dateien aus der CAD-Applikation (bsp. Produktdaten) und der Layout-Applikation (bsp. Arbeitsplätze) im 3dxml-Format exportiert und in ein hochauflösendes visualisierendes Softwaretool der Businessplattform importiert. Dadurch erhält man ein Datenformat zur Darstellung von Hologrammen. Über weitere Anwendungen werden die entsprechenden Objekte über die Datenbrille Microsoft HoloLens projiziert um beispielsweise für eine wesentlich höhere Fertigungsstückzahl eine Inselfertigung in eine Linienfertigung zu wandeln. Unter Beachtung aller Restriktionen der Fabrik und exakter Positionierung der realen Arbeitsplätze aufgrund der Hologramm- Darstellung, die sich aus Berechnungen im Hintergrund auf Basis der Vernetzung aller Daten ergeben, spricht man von „mixed-Reality" und sogenannten „Live-Szenarios".

5 Fazit

Das entwickelte SFM-Konzept stellt ein ganzheitliches System dar, welches die aktuellen industriellen und gesellschaftlichen Trends berücksichtigt und die daraus abgeleiteten Zielsetzungen für das Shopfloor Management der Zukunft erfüllt. Durch das eingesetzte System wird die Transparenz unmittelbar am Ort der Wertschöpfung erhöht und dadurch die Grundlage für die Verlagerung von Entscheidungen zum

Abbildung 7: Positionieren der realen Arbeitsplätze mit der Datenbrille auf Basis der Hologramme und Berechnungen

Shopfloor geschaffen. Durch die Betrachtung von Vergangenheits- und Echtzeitdaten und den daraus abgeleiteten Prognosen wird sowohl die Geschwindigkeit, als auch die Qualität der Entscheidungen signifikant erhöht. Aufgrund der variablen und nutzerspezifischen Darstellung produktionsrelevanter Kennzahlen werden Informationen zielgerichtet zur Verfügung gestellt. Der Mitarbeiter nimmt durch die Einbeziehung seiner individuellen Beanspruchung und der Möglichkeit einer flexiblen Zeiteinteilung einen höheren Stellenwert innerhalb der Produktionsumgebung ein. Das Potenzial der Mitarbeiter kann durch ein einfaches und innovatives Ideenmanagement besser zur kontinuierlichen Verbesserung des Unternehmens genutzt werden. Zusätzlich wird durch die Nutzung einer Cloud-basierten Systemarchitektur die dezentrale Verfügbarkeit aller Informationen und Funktionen gewährleistet.

Zur Weiterentwicklung des entwickelten Systems gilt es, die eingesetzten Berechnungen, Prognosetechniken und Simulationen der Applikationen umfassend in der realitätsnahen Produktionsumgebung der LLF zu validieren und verifizieren. Die Einbeziehung ergänzender Technologien, wie beispielsweise der Einsatz neuer intelligenter Werkzeuge, die Einbindung von Indoor-Lokalisation und VR-Lösungen, ist als Teil der weiteren Forschungstätigkeiten angedacht.

Quellen

Spath, D.; Ganschar, O.; Gerlach, S.; Hämmerle, M.; Krause, T.; Schlund, S.: Produktionsarbeit der Zukunft – Industrie 4.0. Stuttgart: Fraunhofer Verlag. 2013.

Parment, A.: Die Generation Y - Mitarbeiter der Zukunft. Herausforderungen und Erfolgsfaktor für das Personalmanagement. Wiesbaden: Gabler Verlag. 2009.

Diabaté, S.; Bujard, M.: Die Generation Y im Spagat zwischen egalitärem Leitbild und rigiden Strukturen. In: Frankfurter Hefte; 19-22. Frankfurt: Frankfurter Hefte. 2015.

o.V.: Was ist Industrie 4.0? http://www.plattform-i40.de/I40/Navigation/DE/Industrie40/WasIndustrie40/was-ist-industrie-40.html. Abgerufen am

19.01.2017.

Westkämper, E.: Turbulentes Umfeld Von Unternehmen. In: Wandlungsfähige Produktionsunternehmen: Das Stuttgarter Unternehmensmodell; 7-23. Berlin: Springer Verlag. 2009.

Gorecky, D.; Schmitt, M.; Loskyll, M.: Mensch-Maschine-Interaktion im Industrie 4.0-Zeitalter. In: Handbuch Industrie 4.0; Band 4; 2. Auflage; 219-236. Kaiserslautern: Springer Verlag. 2017.

Gröger C.; Stach, C.; Mitschang, B.; Westkämper, E.: A mobile dashboard for analytics-based information provisioning on the shop floor. In: International Journal of Computer Integrated Manufacturing. 2016.

Lendzian, H.; Martin-Martin, R.: Shopfloor-Management: Nachhaltige Problemlösungen schaffen. In: Erfolgsfaktor Lean Management 2.0; 83-98. Berlin Heidelberg: Springer Gabler Verlag. 2016.

McCoy, D.: Business Activity Monitoring: Calm before the Storm. In: Gartner Research Note LE-15-9727. 2002.

Muehlen, z.M.; Shapiro, R.: Business Process Analytics. In: Handbook on Business Process Management. 2. Strategic Alignment. Governance, People and Culture; 137-158. Berlin: Springer Verlag. 2010.

Kletti, J.: Manufacturing Execution Systems - MES. Berlin: Springer Verlag. 2007.

Schumacher, J.; Hummel, V.: Decentralized control of logistic processes in cyber-physical production systems at the example of ESB Logistics Learning Factory. 2016.

Abele, E.; Metternich, J; Tisch, M.; Chryssolouris, G.; Sihn, W.; ElMaraghy H; Hummel, V.; Ranz, F.: Learning Factories for research, education, and training. 2015.

Bracht, U.; Hackenberg, W.; Bierwirth; T.: A Monitoring Approach for the Operative CKD Logistics Increased Efficiency through Optimized Flows of Information by Means of Digital Factory Tools. In; Werkstatttechnik 101 (3): 122–127. 2011.

Schöllhamer, O.: Mehr Effizienz in der Projektarbeit – Shopfloor Management in der Verfahrensentwicklung. Stuttgart. 2013.

Hummel, V.: Industrial Engineering. Modul 6: Belastung, Beanspruchung und Humanisierung der Arbeit. Reutlingen: Hochschule Reutlingen. o.J..

Steven, M.: Hierarchische Produktionsplanung. 2. Auflage. Berlin Heidelberg: Springer Verlag. 2013.

Lieferantenmanagement im Zeitalter der digitalen Transformation

Elisabeth Fröhlich[a], Kristina Steinbiß[b], Matthew Wexford Whelan[c]

[a] *CBS Cologne Business School, D-50667 Köln*
[b] *ESB Business School, Reutlingen University, D-72762 Reutlingen*
[c] *Luton LU1 3QX, England*

Abstract

Die Überlebensfähigkeit von Unternehmen am Markt wird vorrangig durch den Input ihrer Lieferanten getrieben, da dem Einkauf bis zu 50% des prognostizierten Innovationspotenzials zugesprochen wird. Dabei gilt vor allem die digitale Transformation als wesentliche Voraussetzung, um die eigene Wettbewerbsfähigkeit langfristig zu sichern. Der vorliegende Beitrag setzt sich daher mit dem Digitalisierungspotenzial strategischer Lieferantenbeziehungen auseinander. Aufbauend auf einem konzeptionellen Modell des Einkaufs 4.0 werden dessen Erfolgstreiber analysiert. Neben der Optimierung von Einkaufsprozessen durch eProcurementlösungen sollte der Fokus auf der Beziehungspflege zu Lieferanten liegen, um das Potenzial von Einkauf 4.0 vollständig ausschöpfen zu können. Denn nur durch den Aufbau strategischer Wertschöpfungspartnerschaften lässt sich das Erfolgspotenzial des Einkaufs langfristig heben.

1 Die Zukunft des Einkaufs?

1.1 Einkauf 4.0: Bedeutung und Vision

Die zunehmende Individualisierung und Dynamisierung der Absatzmärkte verlangt Agilität und Flexibilität von Unternehmen. Produktentwicklungszyklen sind zu verkürzen, um den stetig steigenden Ansprüchen der Konsumenten gerecht werden zu können. Die Märkte werden damit immer weniger vorhersehbar, der Kosten- und Innovationsdruck im Unternehmen steigt stetig. Gerade auf Seiten des Einkaufs führt dies zunehmend dazu, dass Einkaufsprozesse immer effizienter und damit schlanker werden. Nur so kann der Einkauf auch zukünftig seinen Wertbeitrag zur unternehmerischen Wettbewerbsfähigkeit steigern (Fröhlich und Karlshaus, 2017a, S. 21 ff.).

Durch die digitale Transformation von Unternehmen, die Verwendung digitaler Technologien, sowie die Berücksichtigung deren Auswirkungen, wird sich die Definition von „Wert“ völlig verändern und damit auch der Wertbeitrag des Einkaufs neu bemessen werden. Der Fokus darf dabei zukünftig nicht mehr allein auf der Kostenreduktion (der Generierung von „Savings“) liegen, sondern manifestiert sich vor allem in der Implementierung von Maßnahmen, die zu einer strategischen Differenzierung im Wettbewerb beitragen (Nowosel et al., 2015, S. 5). Die zentrale zukunftsweisende Frage ist, wie der Einkauf sich diesen neuen Herausforderungen stellen kann: Wie können Maßnahmen entwickelt und evaluiert werden, die die unternehmerische Gesamtzielsetzung unterstützen?

Dabei gilt es zu beachten, dass dieser grundlegende strategische Wandel nicht alleine aus der Funktion Beschaffung heraus getrieben werden kann, sondern nur in Kooperation mit den Lieferanten realisierbar ist. Der Lieferant wird zukünftig nicht „nur“ ein weiteres Glied in der Lieferkette sein, sondern als interne Ressource des beschaffenden Unternehmens Innovationen und Digitalisierung voranbringen. Lieferanten als Treiber der Wettbewerbsfähigkeit werden eine Hauptquelle innovativer und auch nachhaltiger Ideen sein. Aktuelle Studien gehen davon aus, dass bis zu 50% aller Innovationen zukünftig aus dem Einkauf kommen (Nowosel et al, 2015, S. 10).

Die Vision, den Einkauf zu einem agilen und innovationstreibenden Impulsgeber zu entwickeln, setzt dabei drei Kernelemente voraus: Neben einem lieferantenorientierten Innovationsmanagement und der digitalen Interaktion in der Supply Chain ist vor allem ein digitales Lieferanten- und Categorymanagement notwendig (Schreiber et al, 2016, S. 2). Deswegen liegt der Fokus der weiteren Ausführungen auf der digitalen Ausgestaltung einer Beschaffer-Lieferanten-Beziehung.

1.2 Voraussetzungen für einen digitalen Einkauf

Um eine Digitalisierungsstrategie im Einkauf etablieren zu können, ist es notwendig, den „Reifegrad“ der Beschaffungsfunktion zu definieren. Aufbauend auf den Ausführungen von Lockamy und McCormack (2004, S. 276) bedarf es der Reifegradstufe 4 - Integration -, um das Innovations- und Agilitätspotenzial einer digitalen Beschaffer-Lieferanten-Beziehung ausschöpfen zu können. Die nachfolgende Diskussion der einzelnen Stufen (siehe Abbildung 1) erläutert die soeben gemachte Aussage.

Im Rahmen der *ersten Reifegradstufe* (*Ad hoc*) lässt sich die Beschaffung als eine rein reaktive Funktion ohne strategische Richtung definieren. Die meiste Zeit wird auf Routinetransaktionen verwandt. Die Einkaufsabteilung hat eine sehr geringe Sichtbarkeit im Unternehmen und ist kaum in mögliche Kooperationen involviert. Lieferanten werden nach Preis und Verfügbarkeit ausgewählt, viele Prozesse sind eher unstrukturiert.

In der *Reifegradstufe 2* (*Unabhängig*) werden aktuelle Beschaffungstrends und -techniken angenommen und der Einkauf agiert als eigenständige, unternehmerische Einheit. Der Fokus liegt hier auf Kosteneffizienz. Die Verbindung zu technischen Unternehmensfunktionen besteht und das Top-Management erkennt den möglichen Wertbeitrag des Einkaufs. Die Supply Chain Management-Prozesse sind definiert und werden dokumentiert.

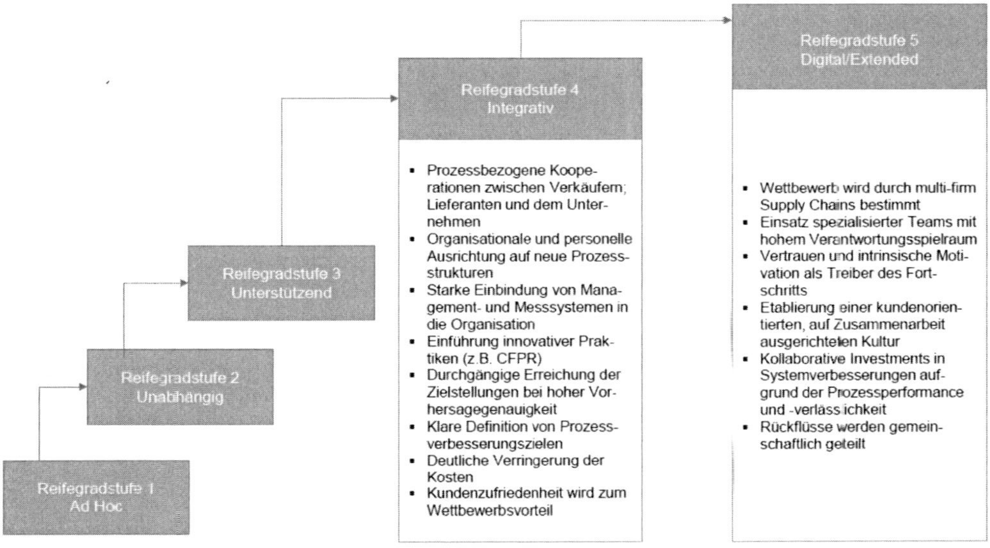

Abbildung 1: Reifegradphasen in der Beschaffung (Höltz, 2012, S. 116)

Unterstützt die Beschaffung die Wettbewerbsstrategie des Unternehmens, spricht man von *Reifegradstufe 3* (*Unterstützend*). Der Einkauf ist zunehmend strategisch ausgerichtet und in Verkaufsprozesse eingebunden. Dabei werden Lieferanten als wertvolle Ressource betrachtet. Eine regelmäßige Analyse von internen (Intralogistik) und externen Beziehungen zu Kunden und Lieferanten findet statt.

In der *Reifegradstufe 4* (*Integrativ*) ist die Beschaffung ein integraler Bestandteil der Unternehmensstrategie und beteiligt sich aktiv an der Formulierung von langfristigen Zielen und Strategien. Bereichsübergreifende, etablierte Kommunikationsnetzwerke mit allen funktionalen Bereichen werden ausgebaut. Die Leistung der Beschaffung wird anhand des Wertbeitrags zum Unternehmenserfolg gemessen.

Spiegelt man diese Charakterisierung der Reifegradstufe 4 an den gemachten Aussagen des Abschnitts 1.1, wird deutlich, dass die Funktion Einkauf die Wettbewerbspotenziale, die sich aus der Etablierung einer digitalen Lieferanten-Beschaffer-Beziehung heben lassen, erst dann ausschöpfen kann, wenn eine integrative Einbindung der Beschaffungsfunktion in die strategische Gesamtausrichtung des Unternehmens stattgefunden hat.

In der *Reifegradstufe 5* (*Digital / Extended*) lebt die Beschaffung das konzeptionelle Modell der Digitalisierung. Dieses wird im folgenden Kapitel detailliert beschrieben.

Wie bereits zuvor ausgeführt, findet sich die Basis zur digitalen Transformation des Lieferantenmanagements in der Reifegradstufe 4. Der Schwerpunkt liegt demnach zunächst in einer klaren Definition von Prozessverbesserungszielen im Kontext einer durchgängigen Fokussierung auf die gesamtunternehmerische Zielstruktur. Somit wirkt der Einkauf - und damit vor allem die Einkaufziele - direkt auf die

Finanzkennzahlen des Unternehmens ein. Diese notwendige Neuausrichtung (Fröhlich und Karlshaus, 2017a, S. 2 ff.) des Einkaufs bedarf folglich auch einer organisationalen und personellen Neuausrichtung sowie einer konsequenten Einbindung in die Management- und Messsysteme der Gesamtorganisation. Letztendlich wird, zur ganzheitlichen Ausrichtung auf die Kundenzufriedenheit als Wettbewerbsvorteil der Zukunft, die Einführung neuer Methoden im Einkauf unumgänglich.

Beispiele zur Ausgestaltung der digitalen Transformation des Lieferantenmanagements im Rahmen der Reifegradstufe 5 werden mithilfe einer „Supplier Journey" in Abschnitt 3 näher erläutert.

2 Ein konzeptionelles Modell der Digitalisierung

Das konzeptionelle Modell der Digitalisierung basiert auf vier grundlegenden „Treibern" der digitalen Transformation. Aufbauend auf der Individualisierung bzw. Hybridisierung von Produkten im Rahmen von *Industrie 4.0* und der damit verbundenen Integration von Kunden und weiteren relevanten Stakeholdern in die Geschäftsprozesse, entsteht im Rahmen der Beschaffung Wertschöpfung vorrangig durch Kollaboration (Schuh et al, 2015, S. 12) und Echtzeitinformationen entlang der gesamten Wertschöpfungskette (Koch et al, 2014, S. 10). Studien gehen davon aus, dass durch den Einsatz von Instrumenten der Digitalisierung in der Funktion Beschaffung u.a. bis zu 40% der Instandhaltungskosten wegfallen, die Zeit bis zum Markteintritt (Time to Market) sich um bis zu 50% reduzieren lässt oder die immensen Datenmengen eine 85% höhere Prognosesicherheit erlauben (McKinsey, 2016, S. 7). Unternehmen können die so gewonnenen Kostensenkungspotenziale direkt an ihre Kunden weitergeben.

Ein weiterer wichtiger „Pusch-Faktor" im Rahmen der digitalen Transformation im Einkauf ist *Big Data*. Die enormen Datenmengen, die dem Einkaufsbereich zur Verfügung stehen, kommen aus den Bereichen Mobilfunk, Finanzindustrie, Energiewirtschaft, Gesundheitswesen und Verkehr oder aus Quellen wie intelligenten Agenten, sozialen Medien, Kredit- und Kundenkarten, Smart-Meter-Systemen, Assistenzgeräten, Überwachungskameras sowie Flug- und Fahrzeugen. Spezielle Lösungen sind notwendig, um diese Datenmengen adäquat speichern, verarbeiten und auswerten zu können. Die zukünftige Wettbewerbsfähigkeit von Unternehmen wird vor allem durch die Professionalität des Umgangs mit den digitalen Datenströmen bestimmt, da dies die Unternehmen dazu befähigt, ihre Kernprozesse besser vorherzusagen und zu kontrollieren (Cavanillas et al, 2016, S. 10). Dabei gilt der Datenstrom entlang des Produktlebenszyklus als eine der am wenigstens genutzten Quelle für Innovation (Bechtold et al, 2016, S. 11). Das Potenzial für den Einkauf ergibt sich vor allem durch die Einbindung in cross-funktionale Produktentwicklungsdatenbanken oder datengetriebene Prozess-Analytics-Aktivitäten (Manyika et al, 2011, S. 78). Der Aufbau einer Big Data Value Chain zur langfristigen Lösung der Probleme in Bezug auf die Datensicherheit ist dabei essentiell.

Neben Industrie 4.0 und Big Data bilden *Cyber-physische Systeme* (*CPS*) und das *Internet der Dinge* (*IoT*) weitere digitale Treiber der Transformation im konzeptionellen Modell der Digitalisierung im Einkauf. CPS sind Systeme mit eingebetteter Software, die über Sensoren und Aktoren verfügen (Esterle und Grosu, 2016, S. 299), erfasste Daten auswerten und speichern, mit Kommunikationseinrichtungen untereinander sowie in globalen Netzen verbunden sind, weltweit erhältliche Daten und Dienste nutzen und über Mensch-Maschine-Schnittstellen verfügen (Gilchrist, 2016, S. 36 f.). SMART Produkte sind dabei ein wesentlicher Output dieser Technologie. Sie sind nicht nur verbunden, intelligent und reaktionsfähig (Kang et al, 2016, S. 111), sondern selbstlernend und passen sich z.B. an besondere Serviceanforderungen an (Bechtold et al, 2016, S. 6 f.). CPS können dazu genutzt werden, Daten aus Produktionslogistik, Kapazitätsplanung und Qualitätssicherung (z.B. notwendige Inspektionen an Maschinen) in Echtzeit miteinander zu verbinden.

IoT hingegen bezeichnet die Vernetzung von Gegenständen mit dem Internet, damit diese Gegenstände selbstständig über das Internet kommunizieren und so verschiedene Aufgaben für den Nutzer wahrnehmen können. Der Anwendungsbereich erstreckt sich dabei von einer allgemeinen Informationsversorgung über automatische Bestellungen bis hin zu Warn- und Notfallfunktionen (Manyika et al, 2015, S. 1). IoT unterstützt Maschine-zu-Maschine-, Mensch-zu-Maschine- und Mensch-mit-Umfeld-Beziehungen (Khodadadi et al, 2016, S. 7) und erleichtert das Life-Cycle Management von Produkten sowie die Steuerung von Kollaborationsprozessen in der gesamten Value Chain (Russo, 2015, S. 9).

Der vierte relevante Treiber der Digitalisierung ist das *Cloud-Computing*. Cloud-Computing beinhaltet Technologien und Geschäftsmodelle, um IT-Ressourcen dynamisch zur Verfügung zu stellen und die Nutzung nach flexiblen Bezahlmodellen abzurechnen. Anstelle des Betreibens von IT-Ressourcen, beispielsweise Servern oder Anwendungen in unternehmenseigenen Rechenzentren, sind diese bedarfsorientiert und flexibel in Form eines dienstleistungsbasierten Geschäftsmodells über das Internet oder ein Intranet verfügbar (Kushida et al, 2011, S. 211). Der wesentliche Vorteil für den Einkauf liegt neben der Reduktion von Hardwarekosten, bei optimaler Ausschöpfung von Rechnerleistung, vor allem in der Einsparung von Lizenzgebühren und Implementierungskosten von Systemen (Narayanan et al, 2017, S. 32). Trotz eventuell begrenzt vorhandener finanzieller Ressourcen erhält der Einkauf Zugang zu relevanten Daten. Dem gegenüber steht jedoch die Tatsache, dass die Daten außerhalb des Unternehmens liegen mit den damit verbundenen Sicherheitsrisiken.

Als theoretisches Rahmenkonzept für die Entwicklung des vorliegenden Modells zur Digitalisierung des Einkaufs wurde der Portfolio-Ansatz von Kraljic (1983) gewählt. Die Grundidee des Portfolios liegt darin, strategisch relevante Beschaffungsobjekte zu identifizieren basierend auf den beiden Dimensionen Wertbeitrag und Beschaffungsrisiko. Verallgemeinert geht es darum, vorhandene Ressourcen bestmöglich gemäß den größten Erfolgsaussichten in unterschiedliche unternehmerische Aktivitätsfelder zu leiten. Digitale Transformation ist „menschen-getrieben": Mitarbeiter sind der Ausgangspunkt für den Einsatz der zuvor beschriebenen „Push-Faktoren" Big Data, CPS und IoT sowie dem Cloud-Computing. Aufbauend auf diesen Faktoren des digitalen Transformations-Push leitet das in Abbildung 2 dargestellte Modell im Geiste Kraljics adäquate Digitalisierungsstrategien in der Einkaufsfunktion gemäß den beiden Dimensionen Beziehungsintensität und Digitalisierungspotenzial ab.

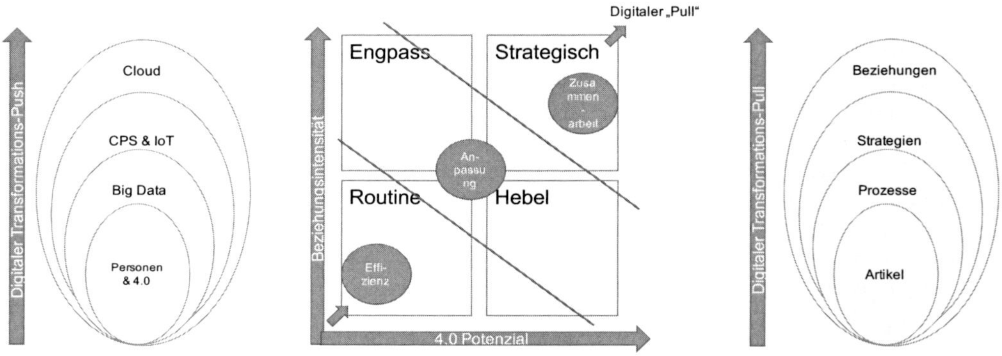

Abbildung 2: Das konzeptionelle Modell der Digitalisierung

Um in diesem Modell effizientes Beschaffungshandeln zu ermöglichen, müssen eProcurement-Lösungen auf Artikel- und Prozessebene im Unternehmen implementiert werden. Dies lässt dann in einem nächsten Schritt die digitale Anpassung der Beschaffungsstrategien zu und entspricht damit der Stufe vier des aufgezeigten Reifegradmodells. Die vollständige digitale Adaption und damit das Erreichen eines Reifegrades der Stufe fünf gelingt den Unternehmen allerdings nur dann, wenn der Fokus zudem auf die Neuausrichtung der Lieferanten-Beschaffer-Beziehungen, der digitalen Zusammenarbeit, liegt. Mögliche Beispiele werden im nächsten Abschnitt näher erläutert.

3 Digitale Lieferanten-Beschaffer-Beziehungen

3.1 Ein geschlossenes Beziehungs-Zyklus Modell

Der vorliegende Beitrag legt den Fokus auf die Etablierung digitaler Lieferanten-Beschaffer-Beziehungen als ein wesentlicher Bestandteil des konzeptionellen Modells des Einkaufs 4.0. Dabei beinhaltet ein strategisches Lieferanten-Management-System neben der Suche nach potenziellen Lieferanten auch die konsequente Bewertung dieser mittels zuvor definierten Kriterien. Ziel ist es, aufbauend auf einer systematischen Leistungsanalyse der Lieferanten langfristige Wertschöpfungspartnerschaften zu etablieren (Fröhlich und Jürgens, 2017, S. 18). Als Basis der weiteren Ausführungen dient das geschlossene Beziehungs-Zyklusmodell von Christopher und Jüttner (2000, S. 119 ff.), welches in Abbildung 3 dargestellt ist.

Hierbei wird die Entwicklung strategischer Partnerschaften auf fünf Kernschritte bezogen. Zunächst gilt es für das Unternehmen zu definieren, mit welchen Lieferanten überhaupt strategische Partnerschaften eingegangen werden sollen. Um in einem zweiten Schritt diese ausgewählten strategischen Beziehungen abbilden zu können, muss das Unternehmen seine Schnittstelleninfrastruktur definieren und in einem dritten Schritt die notwendigen Kommunikationssysteme implementieren. Nur

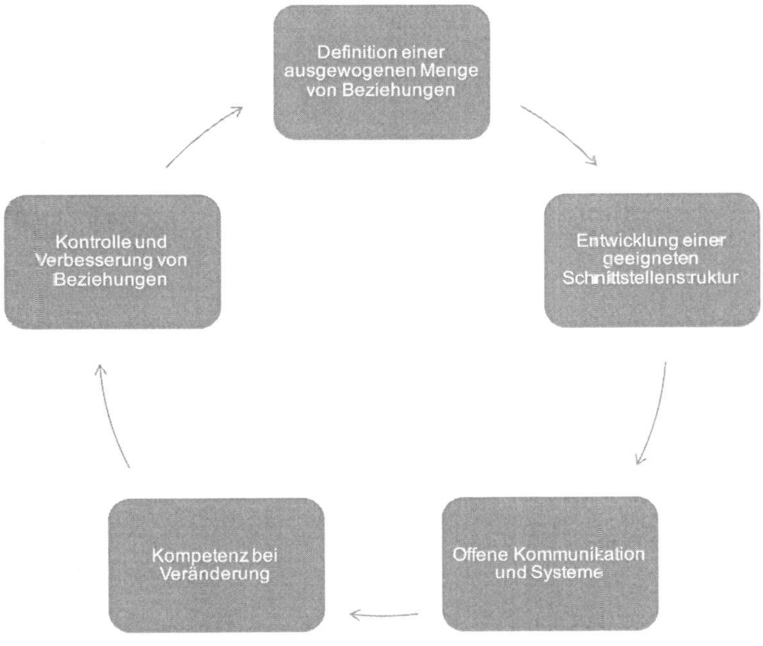

Abbildung 3: Der geschlossene Beziehungszyklus (Christopher und Jüttner, 2000, S. 119)

so lässt sich ein effizienter Informationsaustausch sicherstellen. Die Entwicklung von Partnerschaften bedeutet vor allem, dass Menschen miteinander interagieren. Deswegen müssen Unternehmen die veränderte Beschaffungskultur leben und alle Beteiligten in die Veränderungen mit einbeziehen. Als letzter Schritt ist es wichtig, dass die definierten Prozesse kontrolliert und gegebenenfalls Verbesserungsmaßnahmen entwickelt und umgesetzt werden. Hier gilt es, Transparenz aufzubauen, um langfristig Erfolg zu haben.

Bezieht man diese Überlegungen auf die digitale Transformation, so lassen sich zwei Handlungsebenen (van Echtelt et al, 2008, S. 183) unterscheiden. Auf Ebene der *kurzfristigen Zusammenarbeit* stehen vor allem operative Prozesse des Projekt- und Produktmanagements im Fokus. Das Ziel liegt dabei primär in der Erreichung technischer Leistungsvorteile, Kostenvorteile oder auch kürzeren Vorlaufzeiten. Im Projektmanagement werden Make-or-Buy-Entscheidungen, die Prozesse zur Auswahl neuer Projektpartner und ihr Involvement, die Koordination der Aktivitäten sowie die Bestellung und Nachverfolgung von Prototypen analysiert. Das Produktmanagement stellt dabei auf Informationen über neue Produkte und Technologien ab, auf das Prüfen von Möglichkeiten der Produktstandardisierung und die Evaluierung von Produktdesign in Bezug auf Verfügbarkeit der Teile, Vorlaufzeit, Zyklusdauer, Qualität und Kosten.

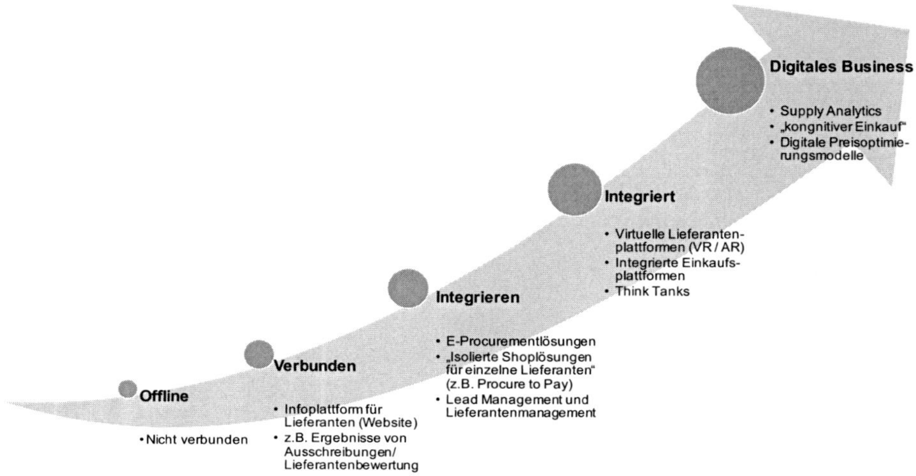

Abbildung 4: Die digitale Supplier Journey

Der Fokus des digitalen Lieferantenmanagements liegt jedoch eindeutig auf der strategischen Handlungsebene mit dem Ziel einer *langfristigen interaktiven Zusammenarbeit*. Diese lässt sich über das Entwicklungsmanagement und Lieferantenschnittstellenmanagement definieren. Dabei gilt es neben der Effizienz auch die Effektivität der zukünftigen Zusammenarbeit zu verbessern, Zugang zu Lieferantentechnologien zu erhalten sowie eine Lösungsdatenbank für zukünftige Projekte zu etablieren. Die gemeinsame langfristige Zielsetzung von Beschaffer und Lieferant sollte die Ausarbeitung eines abgestimmten Technologiefahrplans erlauben.

3.2 Die Supplier Journey

In diesem Kontext von Entwicklungs- und Lieferatenschnittstellenmanagement ist folgende beispielhafte Darstellung der Supplier Journey zu verstehen. In Anlehnung an das empirisch bereits besser belegte Konzept der „Customer Journey" kann die folgende „Supplier Journey" (Abbildung 4) als theoretisches Modell interpretiert werden, um die vielschichtigen Abläufe und Touchpoints in einer Lieferanten-Beschaffer-Beziehung vor dem Hintergrund der digitalen Transformation neu zu definieren und damit innovative Wege für die Funktion Beschaffung aufzuzeigen (de Salles Canfield und Basso, 2017, S. 105).

Greift man auf die zuvor gemachten Ausführungen des Reifegradmodells in Abschnitt 1.1 zurück, beginnt die „Supplier Journey" gemäß Reifegradstufe 1 mit einer traditionellen „offline"-Beschaffer-Lieferanten-Beziehung. Erste Schritte in Richtung Unabhängigkeit lassen sich realisieren, indem man dem Lieferanten bewusst an die

Einkaufsfunktion im beschaffenden Unternehmen anbindet. Klassische Informationsplattformen auf der Website des Beschaffers erlauben es dem Lieferanten, Ergebnisse von Ausschreibungen oder Bewertungsprozessen einzusehen, um daraus zu lernen und gemeinsame Verbesserungspotenziale zu identifizieren und zu realisieren. Reifegradstufe 3 verlangt nach integrierenden Maßnahmen, um die unterstützende Funktion des Einkaufs auch in einer Lieferanten-Beschaffer-Beziehung implementieren zu können. eProcurementlösungen sind in dieser Phase der Supplier Journey der zentrale Ansatz, bilden aber auch in dem in Abschnitt 2 dargestellten konzeptionellen Modell die Eingangsstrategie der digitalen Transformation im Einkauf ab. Elektronische Beschaffungslösungen weisen jedoch einen sehr geringen Grad an Beziehungsintensität und Digitalisierungspotenzial auf. Eine Integration von netzwerkbasierten Kommunikations- und Informationstechnologien ermöglicht dabei den elektronischen Einkauf von Beschaffungsobjekten. Eine adäquate Lieferanteneinbindung spielt an dieser Stelle eine zentrale Rolle (Kollmann, 2016, S. 121 f.). Isolierte Shoplösungen für Lieferanten, z.B. Procure to Pay Lösungen, können ein erster Schritt zur integrierten eProcurement Lösung sein.

Um eine durchgängige Realisierung der Ziele bei hoher Vorhersagegenauigkeit zur deutlichen Verringerung der Beschaffungskosten, aber auch der Verbesserung der Endkundenzufriedenheit als qualitativen Wettbewerbsvorteil zu ermöglichen, sind in der Reifegradstufe 4 (vergleiche Abbildung 1) prozessbezogene Kooperationen zwischen Beschaffer und Lieferant zu etablieren. Greift man auf die Erläuterungen bezüglich der Treiber im konzeptionellen Modell der Digitalisierung zurück, können virtuelle Lieferantenplattformen oder integrierte Einkaufsplattformen in dieser Phase der Supplier Journey zum Einsatz kommen. Im Sinne eines Think Tank können sogenannte „Virtuelle Lieferanten Räume“ aufgebaut werden. Diese virtuelle Interaktionsplattform erlaubt es Beschaffern, sich mit ihren Lieferanten global zu relevanten Supply Chain Themen auszutauschen, gemeinsam Ideen und Innovationen zu entwickeln. Interessant könnte auch sein, dass noch nicht gelistete Lieferanten Einblick in die Bedürfnisstruktur des Beschaffers erhalten und somit potentielle, innovative Lieferquellen dem beschaffenden Unternehmen zugänglich gemacht werden (Schreiber et al, 2016, S. 6). Think Tanks in Beschaffer-Lieferanten-Beziehung finden gerade erst Einzug in strategisches Beschaffungshandeln. Noch handelt es sich dabei um face-to-face Formate, wo sich das beschaffende Unternehmen in Co-working Spaces oder Innovationszentren zum gemeinsamen Austausch zusammenfindet.

Um in der letzten Stufe der Supplier Journey eine auf den Endkunden, auf Kollaboration ausgerichtete Multi-Unternehmens-Lieferkette implementieren zu können, bedarf es einer vertrauensvollen Beschaffer-Lieferanten-Beziehung, gestützt durch relevante und gesicherte Daten. Big Data spielt in diesem Kontext eine große Rolle und unterstützt die Entscheidungsfindungen in einer strategischen Wertschöpfungspartnerschaft. Eine konkrete Anwendung in diesem Kontext könnten „Supply Analytics“ sein, die sich im Sinne eines „kognitiven Einkaufs“ so entwickeln werden, dass sie valide Informationen zur Entscheidungsunterstützung zur Verfügung stellen und dem Einkauf Möglichkeiten aufzeigen, wie Einkaufsprozesse weiter optimiert werden können (Schreiber et al, 2016, S. 6). Digitale Preisoptimierungsmodelle, die z.B. bereits im Bereich Energieeinkauf existieren, können als konkretes Anwendungsbeispiel genannt werden.

4 Die Zukunft des Einkaufs!

„Der Einkauf schrumpft - der operative Einkauf wird weitgehend automatisiert“ heißt es in einer aktuellen Studie zum Thema Einkauf 4.0 des Fraunhofer-Instituts in Zusammenarbeit mit dem BME e.V. (Henke und Feldmann, 2016, S. 8). Die Aussage, dass es den klassischen Einkauf in der Zukunft nicht mehr geben wird, lässt sich durch die zuvor gemachten Überlegungen in diesem Beitrag eindeutig herleiten. Elektronische Einkaufsprozesse automatisieren die operative Einkaufsfunktion, es ist lediglich eine Überwachung dieser Prozesse durch den strategischen Einkauf in der Zukunft notwendig. Über den Erfolg eines Unternehmens wird zukünftig nicht mehr das Generieren von Savings im Sinne einer kontinuierlichen Optimierung der Beschaffungsprozesse entscheiden, sondern inwieweit es dem beschaffenden Unternehmen gelingt in virtuellen „Company Malls“ (Schreiber et al, 2016, S. 6) den Lieferanten als weitere Ressource für Fortschritt und Innovation zu integrieren.

Einkaufsverantwortliche stehen vor der Wahl: Entweder ignorieren sie die immensen Potenziale der Digitalisierung des Einkaufs und riskieren damit die weitestgehende Auslagerung des Einkaufs aus den Kernfunktionen eines Unternehmens. Oder sie öffnen sich den Möglichkeiten der digitalen Transformation und versuchen etwa durch IoT-Anwendungen, den Einkauf neu aufzustellen, um vor allem gemeinsam mit Lieferanten Innovationen und Ideen für die Zukunft der Wirtschaft zu entwickeln.

Häufig findet sich die Aussage, dass man für die Bewältigung der Herausforderungen des Einkaufs 4.0 "ein völlig neues Kompetenzverständnis des Beschaffungsmitarbeiters“ benötigt. Nach Ansicht der Autoren reicht es aus, sich der Verantwortung einer Etablierung eines innovativen Personalentwicklungskonzepts zu stellen. „Virtuelle Führungsmodelle durch die zunehmend globalisierten Einkaufsorganisationen, in denen oftmals weltweite Einkaufsteams an unterschiedlichen Standorten zu managen sind, dienen als markantes Beispiel dafür, dass Agilität und Empathie zunächst in der Funktion Einkauf selbst gelebt werden müssen, um dann die eigenen Mitarbeiter mit auf diese Reise in die neue digitale Zukunft des Einkaufs zu nehmen“ (Fröhlich und Karlshaus 2017).

Literatur

Bechtold, J., Kern, A., Laurenstein, C., Bernhofer, L. (2016): Industry 4.0 - The Capgemini Consulting View: Sharpening the Picture beyond the Hype. Erschienen unter: https://www.capgemini-consulting.com/ resource-file-access/resource/pdf/capgemini-consulting-industrie-4.0.pdf, 25.11.2017

Cavanillas, J. M., Curry, E., Wahlster, W. (2016): The Big Data Value Opportunity, in: New Horizons for a Data-Driven Economy, hrsg. von J. M. Cavanillas, E. Curry

/ W. Wahster, S. 3-11

Christopher, M., Jüttner, U. (2000): Developing Strategic Partnerships in the Supply Chain: A Practisioner Perspective, in: European Journal of Purchasing & Supply Management, Jhg. 06, Nr. 2, S. 117-127

De Salles Canfield, D., Basso, K. (2017): Integrating Satisfaction and Cultural Background in the Customer Journey: A Method Development and Test, in: Journal of International Consumer Marketing, Jhg. 29, Nr. 2, S. 104-117

Esterle, L., Grosu, R. (2016): Cyber-physical systems: Challenge of the 21st Century, in: Elektrotechnik und Informationstechnik, 133. Jhg., Heft 7, S. 299-303

Fröhlich, E., Karlshaus, A. (2017): Digitale Transformation in der Beschaffung, erschienen unter: https://www.springerprofessional.de/industrie-4-0/-corporate-social-responsibility/digitale-transformation-in-der-beschaffung/12234584, 30.05.2017.

Fröhlich, E., Karlshaus, A. (2017a): Status Quo: Personalentwicklung in der Beschaffung - Ergebnisse einer empirischen Studie, in: Personalentwicklung in der Beschaffung, hrsg. von E. Fröhlich, A. Karlshaus, Wiesbaden, S. 1-25

Fröhlich, L., Jürgens, C. (2017): Nachhaltiges Lieferantenmanagement: Notwendigkeit oder Illusion?, in: Der Betriebswirt, Jhg. 58, Nr. 1, S. 15-20

Gilchrist, A. (2016): Industry 4.0. The Industrial Internet of Things, New York

Henke, M., Feldmann, C. (2016): Einkauf 4.0 - Digitalisierung im Einkauf, hrsg. von Fraunhofer-Institut für Materialfluss und Logistik, BME Bundesverband Materialwirtschaft und Einkauf e.V., Frankfurt

Höltz, N. (2012): Lean Logistics Maturity Model, Dissertation, Berlin

Kang, H., Lee, J., Choi, S., Kim, H., Park, J., Son, J., Kim, B., Buckley, N. (2016): Smart Manufacturing: Past Research, Present Findings, and Future Directions, in: International Journey of Precision Engineering and Manufacturing Green Technolgoy, Jhg. 03, Nr. 01, S. 111-128

Khodadadi, F., Dastjerdi, A. V., Buyya, R. (2016): Internet of Things: An Overview, in: Internet of Things - Principles and Paradigms, hrsg. von R. Buyya, A. V. Dastjerdi S. 3-28

Koch, V., Kuge, S., Geissbauer, R., Schrauf, S. (2014): Industry 4.0 opportunities and challanges of the industrial internet, erschienen unter: http://www.strategyand.pwc.com/media/file/industry-4-0.pdf, 26.01.2017

Kollmann, T. (2016): E-Business, 6. überarbeitete Aufl., Wiesbaden

Kraljic, P. (1983): Purchasing Must Become Supply Management, in: Harvard Business Review, Jgh. 61, Nr. 05, S. 109-117

Kushida, K. E., Kurray, J., Zysman, J. (2011): Diffusing the Cloud: Cloud Computing and Implications for Public Policy, in: Journal of Industy, Competitino and Trade,

Jhg. 03, Nr. 11, S. 209-237

Lockamy III, A., McCormack, K. (2004): The development of a supply chain management process maturity model using the concepts of business process orientation, Supply Chain Management: An International Journal, Jhg. 09, Nr. 04, S.272-278

Manyika, J., Chui, M., Brown, B., Bughin, J., Dobbs, R., Roxburgh, C., Hung Byers, A. (2011): Big data: The next frontier for innovation, competition, and productivity, McKinsey Global Institute, Mai 2011

McKinsey & Company (2016): Industry 4.0 at McKinsey´s model factories. Get ready for the disruptive wave, erschienen unter: https://operations-excellence.mckinsey.com/files/downloads/2016/digital40model factories brochure1.pdf, 14.02.2017

Narayanan, S., Sambamurthy, V., Choi, T. Y. (2017): Getting Value from the Cloud, in: Supply Chain Management Review, Jhg. 56, Nr. 02, S. 32-37

Nowosel, K., Terrill, A., Timmermann, K. (2015): Procurement next Frontier - The Future Will Give Rise to an Organization of One, Accenture Strategy

Russo, G., Marsigalia, B., Evangeliste, F., Palmaccio, M., Maggioni, M. (2015): Exploring regulatios and scope of the Internet of Things in contemporary companies: A fist literature analysis, in: Journal of Innovation and Entrepreneuership, Jhg. 11, Nr. 04, S. 1-13

Schreiber, B., Janssen, R., Weaver S., Peinter, S. (2016): Procurement 4.0 in the digital world, Viewpoint Arthur D. Little

Schuh, G., Reuter, C., Hauptvogel, A., Doelle, C. (2015): Hypothesis for a Theory of Production in the Context of Industrie 4.0., in: Advances in Production Technology, hrsg. von C. Brecher, S. 11-23

van Echtelt, F. E. A., Wynstra, F., van Weele, A. J., Duysters, G. (2008): Managing Supplier Involvement in New Product Development: A Multiple Case-Study, in: Journal of Product Innovation Management, Jhg. 25, Nr. 02, S. 180-201

Vom Multivariantengeschäftsmodell zur softwaregesteuerten Individualisierung im Auto

Matthias Geiger[a], Kristina Steinbiß[a]
[a] *ESB Business School, Reutlingen University, D-72762 Reutlingen*

Abstract

Die Welt dreht sich immer schneller. Etablierte Firmen, die sinnbildlich für innovative Produkte stehen, werden in kurzer Zeit vom Markt gedrängt und durch disruptive Ansätze aus ihrem Geschäft verdrängt. Inkrementelle Ansätze zur Verbesserung der Produktqualität, zur Reduktion von Herstellkosten und der Fokus auf die Produktionsoptimierung nach dem Individualisierungsprinzip (Losgröße 1) sind keine auf die Zukunft ausgerichteten Zielzustände für Automobilhersteller mehr. Neue Wettbewerber, seien es Softwaregiganten oder agile Startups, treten in den Markt ein. Die traditionelle Massenproduktion des Automotivbereichs kann den sich stetig verändern Kundenanforderungen nicht mehr länger gerecht werden. Dazu bedarf es Geschäftsmodell-Innovationen, wie beispielsweise digitale Preismodelle oder On-Demand Funktionalitäten.

1 Analyse der Ausgangssituation

1.1 Marktentwicklungen

Die Automobilbranche muss Wandel als Chance begreifen. Vor allem die Entwicklungen im technologischen, demographischen und auch sozialen Bereich erfordern eine Neuausrichtung des klassischen Geschäftsmodells der Automobilindustrie, das bisher vorrangig auf Verkaufen ausgerichtet ist (Stricker et al., 2011, S. 5). Dabei bildet ein Geschäftsmodell die Grundlogik eines jeden Unternehmens. Es beschreibt, welcher Nutzen auf welche Weise für Kunden und Partner gestiftet wird (Schallmo 2013, S. 22 f.). Der Fokus der hier betrachteten Geschäftsmodell-Innovationen liegt jedoch nicht darin, Bestehendes effizienter oder besser zu machen, sondern bewusst das vorhandene Verkaufsmodell der Automobilindustrie in Frage zu stellen und zu verändern. Dabei helfen Geschäftsmodell-Innovationen den Unternehmen vor allem darin, sich gegenüber ihren Wettbewerbern zu differenzieren. Dieses soll im Rahmen der vorliegenden Ausarbeitung durch die stetige Transformation als innovatives, auf

Kundennutzen ausgelegtes und komplett individualisierbares Geschäftsmodell diskutiert werden.

Flatrate-Modelle wie Spotify und Netflix, Freemium Apps und on-demand Services verändern die Wahrnehmung von Produktbesitz und Verfügbarkeit. Der Kunde der Zukunft wird es gewohnt sein, Produkte nicht mehr zu besitzen, sondern sie als Service zu nutzen. Daher müssen zukunftsfähige Produkte an die sich ständig verändernden Lebenssituationen anpassbar sein. So werden beispielsweise Anschaffungen in Sekundenbruchteilen auf dem Smartphone getätigt und Plattformen dominieren die Interaktion mit dem Kunden. Diese Ansprüche gilt es auf das Konzept der Mobilität zu übertragen und mit Hilfe der Analyse von Megatrends zu validieren (Ruff 2012, S. 219).

1.2 Makroökonomische Betrachtung

Die PESTLE-Analyse (Political, Economic, Social, Technological, Legal, Environmental) analysiert die verschiedenen makroökonomischen Einflussfaktoren auf ein Unternehmen bzw. einen Geschäftsbereich, schätzt Umweltbedingungen ab und leitet daraus Handlungsempfehlungen ab (Diez 2015, S. 55). Die Umweltanalyse mit der PESTLE-Methode eignet sich vor allem für die Erfassung eines breiten Spektrums an Einflussfaktoren und dient als Basis für die vorliegende Untersuchung. Durch eine ganzheitliche agile Betrachtung der unternehmerischen Umweltfaktoren lassen sich damit strategische Entscheidungen fundieren (Kerth et al., 2011, S. 118). Für die Automobilindustrie ergibt sich aus der PESTLE-Analyse eine Vielzahl an Einflussfaktoren. In Abbildung 1 sind die Analyseergebnisse der makroökonomischen Faktoren dargestellt (Kerth et al., 2011, S. 122 f.; Wedeniwski 2015, S. 260 ff.).

Deutlich wird, dass alle Umweltbereiche einen enormen Einfluss auf die zukünftige Entwicklung der Automobilindustrie haben. Von strategischer Bedeutung zur Ausrichtung des Geschäfts auf veränderte Kundenanforderungen werden neben der E-Mobilität vor allem die Verschiebung des Absatzmarktes in Richtung Osten sowie das autonome Fahren sein.

1.3 Branchenstrukturanalyse

Bildet PESTLE den makroökonomischen Ausgangspunkt einer holistischen Betrachtung der Einflussfaktoren auf zukünftige Geschäftsmodelle der Automobilindustrie, dient das Branchenstrukturmodell (auch Five-Forces-Modell genannt) von Porter zur Bestimmung der Attraktivität einer Branche (Porter 1989, S. 133). Diese lässt sich auf Grundlage von Marktentwicklung und Wettbewerbssituation als langfristig profitables Entwicklungspotenzial systematisch ableiten. So liefert z.B. die Konkurrenzstärke als eine Attraktivitätsdimension Aussagen zur prognostizierten Gewinnentwicklung (Rusnjak 2014, S. 52). Der Erkenntnisgewinn aus der Branchenstrukturanalyse ist deswegen von erheblicher Bedeutung, da mit ihm Maßnahmen zur Reduktion der Wettbewerbseinflüsse auf das eigene Unternehmen entwickelt werden können.

Ferner dient das Five-Forces-Modell im Rahmen der strategischen Unternehmensanalyse als Konzept zur Antizipation von Dynamiken innerhalb des Marktumfeldes

Politische Einflussfaktoren	Wirtschaftliche Einflussfaktoren	Technologische Einflussfaktoren
• Strengere Regularien für Abgasregelungen • Staatliche Förderung E-Mobilität • Fahrverbote in Innenstädten • Freihandelsabkommen • Ausbau des ÖPNV	• Global Steigender Wohlstand • Wechselkursrisiken • Zinsänderungen • Abnahme von Verkaufszahlen • Verschiebung der Absatzmärkte Richtung Osten	• E-Mobilität • Autonomes Fahren • Connected Cars • Neue Transportlösungen (Lufttaxis etc.) • Digitalisierung • E-Hailing
Soziale Einflussfaktoren	**Rechtliche Einflussfaktoren**	**Umweltbedingte Einflussfaktoren**
• Auto als erweiterter Lebensraum • Carsharing bzw. Shared-Economy • Trend zur Individualisierung • Nachhaltigkeitsgedanke • Besitz eines Automobils verliert an Prestige • Urbanisierung	• Erhöhung von Steuern für Nutzung / Besitz eines Automobils • Mindestquoten für E-Mobilität • Einführung globaler Standards • Freigabe für das autonome Fahren	• Verstärktes Umweltbewusstsein • Fokussierung auf nachhaltige Mobilität • Umweltplaketten • Emissionsregelungen • Recycling • Einfluss von Nachhaltigkeit als Verkaufsargument

Abbildung 1: PESTLE-Analyse der Automobilindustrie

(Schawel et al., 2014, S. 108). Die fünf Dimensionen sind: *potenzielle neue Konkurrenten, Verhandlungsmacht der Lieferanten, Verhandlungsmacht der Kunden, Rivalität zwischen den Wettbewerbern* und *Bedrohung durch Ersatzprodukte*. Eine automobilspezifische Branchenstrukturanalyse ist in Abbildung 2 dargestellt.

Durch Ableitung relevanter Markttrends aus der PESTLE- und der Branchenstrukturanalyse in Verbindung mit den aktuellen Zukunftstrends lassen sich folgende Megatrends für die Zukunft der Automobilindustrie erkennen (Berret, 2016; Clarke und Mullancy, 2015; Ifmo 2010; Schade et al., 2012; Winterhoff, 2009):

- E-Mobilität
- Autonomes Fahren
- Demographischer Wandel (Generation Y, Digital Natives)
- Urbanisierung
- Shared Economy
- Generation Startup
- Marktdurchdringung von Connected Cars

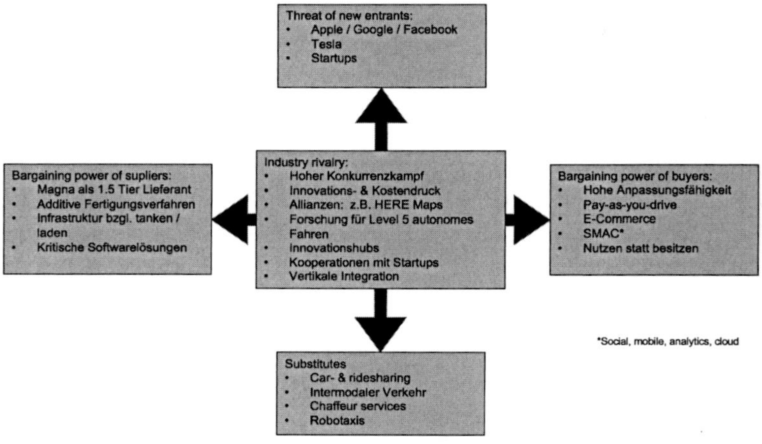

Abbildung 2: Branchenstrukturanalyse der Automobilindustrie

- Auto als erweiterter Lebensraum
- Kundenwunsch nach Individualisierung

1.4 Notwendigkeit zur Veränderung

Die dargestellten Trends sind mit der Fortschreibung des bisherigen Geschäftsmodells in der Automobilindustrie nicht gewinnmaximierend nutzbar. Obwohl die voranstehende Analyse keine lückenlos detaillierte Betrachtung der Ausgangssituation der Automobilindustrie ist, weist sie jedoch ihre wichtigsten Entwicklungen und Herausforderungen auf. Die Notwendigkeit für ein Umdenken ist dementsprechend eindeutig.

Aber nicht nur in der Automobilindustrie, sondern auch übergreifend auf andere Industriezweige ist ein Wandel innerhalb der Geschäftswelt zu erkennen. Eine Betrachtung des Zeitraums der Jahre 2000 bis 2014 zeigt, dass 52 % der Fortune 500 Unternehmen insolvent gegangen sind, aufgekauft worden oder nicht mehr gelistet werden (Didier et al., 2015, S. 78). Dabei ist die Digitalisierung ein wichtiger Treiber neuartiger Ideen und disruptiver Innovationen, die die vorhandene Marktstruktur durchdringt. Basierend auf geschäftsmodellbezogenen Innovationen verdrängen neue Konkurrenten vorherrschende Unternehmen vom Markt. Dabei sind AirBnB und Uber nur zwei der populärsten Beispiele für solch eine Art der Disruption. Diese plattformgetriebenen Lösungen decken Kundenbedürfnisse, die von den etablierten Unternehmen nicht bedient wurden. In der Automobilindustrie ist mit solch einem Trend in den nächsten Jahren zu rechnen. Anhand der vorher beschriebenen Entwicklungen innerhalb des Mobilitätsgeschäfts, Car Sharing, autonomes Fahren, etc., stellt sich die zentrale Frage für den heutigen Automobilhersteller: Wie kann das eigene Unternehmen zukünftig konkurrenzfähig bleiben?

2 Softwaregesteuerte Individualisierung im Auto der Zukunft

Der Begriff des Geschäftsmodells, des Business Models, ist ein in der wissenschaftlichen Literatur nicht genau definierter Begriff (Wirtz 2013, S. 12). Geprägt wurde dieser Begriff vor allem seit dem Aufkommen des Internets und der New Economy (Osterwalder 2004, S. 23). Die unterschiedlichen Modelle innerhalb der Literatur sind einer Analyse auf deren Anwendung für das erdachte Geschäftsmodell analysiert worden. Die Geschäftsmodelle von Schallmo, der Ansatz von Becker und Ulrich, Timmons und Gassmanns Modell und weitere sind auf deren Anwendungsmöglichkeit untersucht worden (Schallmo 2013, S. 52; Becker, Ulrich 2013, S. 13; Gassmann, Frankenberger, Csik 2013, S. 5 ff.). Der Entschluss für die Entwicklung des Geschäftsmodells mit dem Ansatz von Osterwalder ist anhand folgender Vorteile gefallen (Weinreich 2016, S. 202):

- Leichte Verständlichkeit des Grundgerüstes
- Reduktion der Komplexität der Sachverhalte
- Praxisorientierter Ansatz mit schneller Anpassungsmöglichkeit
- Respektierung von Chancen und Risiken
- Erleichterte Kommunikation der wichtigsten Aspekte

Ein Geschäftsmodell definiert sich laut Osterwalder, der das Business Model Canvas entwickelt hat, dadurch, dass Grundprinzipien für eine Organisation beschrieben werden, mit denen diese Organisation Wert generiert, vermittelt und erfasst (Osterwalder et al., 2011, S. 19). Das grundlegende Gerüst, das hinter dem Business Model Canvas steht, ist in Tabelle 1 dargestellt (Kapteyn 2012, S. 3). Das **WIE** beschreibt die Wertschöpfungsarchitektur, das **WAS** beschreibt das Wertangebot, mit **WER** sind die Kunden gemeint und das **WARUM** beantwortet die Fragen nach den Kosten und Erlösen.

Zur Verdeutlichung des Geschäftsmodells soll ein einfaches Beispiel dienen: Automobilhersteller unterscheiden ihre Modellreihen u.a. anhand von Karosserieformen und verbauter Motorleistung. Abbildung 3 zeigt am Beispiel der Motorleistung das Potential eines neuartigen Geschäftsmodells.

Die Fahrzeugklassen 1, 2 und 3 stellen dabei Fahrzeugmodelle der selben Modellreihe mit unterschiedlicher Motorisierung dar. Das Potential, das sich mit der softwaregesteuerten Individualisierung ergibt, ist in der Abbildung gestreift dargestellt. Dieses kleine Beispiel verdeutlicht, dass eine Fokussierung der Produktion auf Fahrzeugklasse 3 mit einer softwaregesteuerten Individualisierung sinnvoll sein kann.

Wie? (Wertschöpfungs-architektur)	Was? (Value Proposition)	Wer? (Kunden)
Warum? (Kosten/Erträge)		

Tabelle 1: Grundstruktur des Business Model Canvas

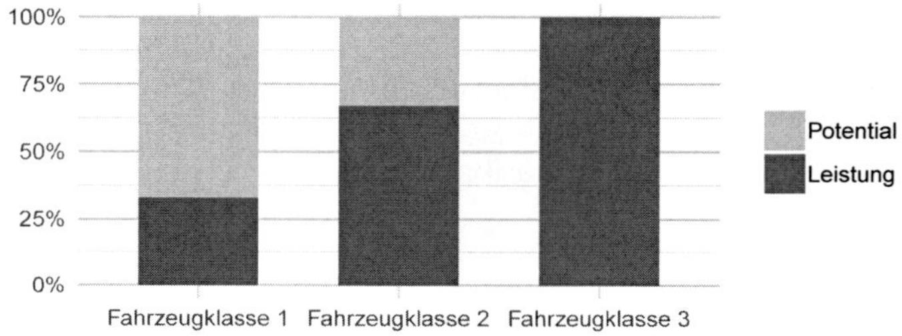

Abbildung 3: Aufgabe des Laststufenkonzeptes und Reduzierung der Varianten

Diese Komplexitätsreduktion gewährleistet dabei sowohl "down-grades" als auch eine volle Nutzung der Motorleistung. Auch wenn dieser Ansatz bezogen auf einen klassischen Verbrennungsmotor als sehr aufwendig zu betrachten ist, ist dieses Vorgehen bezogen auf die Elektromobilität jedoch folgerichtig. Hier können über die softwaregesteuerte Individualisierung die Elektromotoren individuell angesprochen und das Drehmoment verändert werden. Die Differenzierung innerhalb der Modellreihe findet dann anhand der Aktivierung unterschiedlicher Module, z.B. Sitzheizung, Navigation etc., statt und ermöglicht eine individuelle Ansprache der Kundensegmente.

Das Business Model Canvas dient der Beschreibung des Geschäftsmodells einer softwaregesteuerten Individualisierung im Auto der Zukunft. Im Baustein ***Customer Segments*** geht es vor allem darum, den Kundenwunsch nach Individualisierung zu identifizieren. Viele Kunden möchten ein erschwingliches Premiumfahrzeug fahren (Rumpelt 2017, S. 7 ff.). Flexible Anpassungsmöglichkeiten an die Lebenssituation sind dabei genauso wichtig. Die softwaregesteuerte Zubuchung von Produkteigenschaften innerhalb des Automobils soll komfortable sein. Nachrüstbare und On-Demand Lösungen werden von den Kunden heutzutage erwartet.

Value Proposition bedient sich der Individualisierung als Schlüssel, um in Zeiten verschwindend geringer Gewinnmargen sich gegenüber Konkurrenten durchzusetzen und am Markt einen Wettbewerbsvorteil zu erzielen. Wird aktuell die Individualisierung vorrangig durch das Angebot unterschiedlicher Modellreihen und kundenspezifischer Konfigurationen erreicht, so bieten sich durch das Geschäftsmodell völlig neue Ansätze (Stüber 2017, abgerufen am 05.12.2017): Mit der softwaregesteuerten Individualisierung ist eine Möglichkeit für den Kunden gegeben, sein Automobil unabhängig von Händlern, Servicestationen etc. anzupassen. Durch diese ungebundene, ob lokal oder zeitlich gesehene, Buchungsmöglichkeit erlebt der Kunde ein neues Gefühl der Bedürfniserfüllung. Verschiedene Bezahlmöglichkeiten sind innerhalb des Geschäftsmodells verankert. Mit neu aufkommenden Bezahlmodellen wie: Freemium, In-Car-Käufen und Flatrates, sind die Möglichkeiten mannigfaltig. Dem Kunden ist das Wertangebot der vollen Kontrolle über Leistung und Budget angeboten. Mit einer Art von „Performance auf Abruf" ist dem Kunden eine neue Art angeboten, sein Fahrerlebnis individuell zu gestalten. Jeder Kunde bucht die Features, die er für wünschenswert hält, fährt ein an sich angepasstes Auto und ist damit in seinen Ansprüchen an das Fahrzeug erfüllt.

Der Baustein *Channels* beschreibt die unterschiedlichen Kundenkontaktpunkte, wie ein Unternehmen seine Kunden erreicht. Im Rahmen der Digitalisierung ist der Internetauftritt eines Unternehmens in seiner Bedeutung stark gestiegen (Diez 2012, S. 281). Doch auch das Angebot und die damit verbundene Nutzung von temporären Besitzverhältnissen wie Leasing, Sharing, stellen wichtige Kundenkontaktpunkte dar. Die wichtigsten Kanäle sind in der folgenden Auflistung enthalten:

- Händlernetzwerke oder Eigene Vertriebsorganisation
- Direktverkauf über werkseigene Niederlassungen
- Ausstellungsraum
- Internetauftritt mit Plattform
- Leasing / Carsharing

Der Erfolg eines Unternehmens wird maßgeblich durch seine Kundenbeziehungen bestimmt. Der Baustein *Customer Relationships* definiert diese Beziehungen und deren Integration innerhalb des Geschäftsmodells. Aus dem Customer Relationship Management ist ein eigener Unternehmensbereich entstanden, dem eine kritische Rolle zuteil wird (Weinreich 2016, S. 218). Für das Unternehmen ergeben sich unterschiedliche Möglichkeiten der Ausgestaltung seiner Kundenbeziehungen. Sie findet durch persönliche, computerunterstütze oder automatisierte Unterstützung statt. Mit Hilfe von Communitys, Social Media, Big Data Analysen oder anderen Bereichen ist eine zielgruppenorientierte Ausrichtung und Betreuung möglich.

Die Kundenbeziehungen für das Geschäftsmodell mit der softwaregesteuerten Individualisierung teilen sich in zwei verschiedene Bereiche auf: Der erste Bereich behandelt die Kundenbeziehung vor der Kaufentscheidung und der zweite Bereich bezieht sich auf die Kundenbeziehung in der Nutzungsphase des Produktes, also nach der Kaufentscheidung. Vor der Kaufentscheidung ist es für den Kunden wichtig, dass er ausführliche Informationen zu dem Produkt erhält. Der Kunde vergleicht Produkte anhand objektiver Merkmale im Internet, schaut sich nach dem besten Preis um und wägt Alternativen ab (Endlein 2017, S. 16). Eine dementsprechend aufgebaute Informationsplattform ist für die Nutzung des Geschäftsmodells von großer Bedeutung. Die User-Experience und Anwendungsfreundlichkeit sind Kriterien, die dabei im Vordergrund stehen. Eine auf diese Bedürfnisse ausgerichtete Informationsplattform ist die erste Möglichkeit, Einflussnahme auf den Kunden auszuüben. Eine weitere Art der Kundenbeziehung kann eine probeweise Art der Nutzung sein. Leasing und vor allem Carsharing ist hierbei eine denkbare Art, Kunden von den eigenen Produkteigenschaften zu überzeugen und eine erste Beziehung nachhaltig zu entwickeln. Neben der Informationsplattform und dem Leasing bzw. Sharing ist weiterhin der persönliche Kontakt essentiell für das Geschäftsmodell. Digitalisierte Produkte sind für eine Vielzahl an Kunden in ihren Möglichkeiten nicht ersichtlich. Kunden erwarten Beratung, persönliche Tests und ein Eingehen auf Bedürfnisse (Endlein 2017, S. 42). Eine wichtige Aufgabe für Unternehmen besteht zukünftig vor allem darin, die denkbare anfängliche Skepsis der Kunden gegenüber dem Produkt abzubauen.

Die Kundenbeziehung, die nach Kaufentscheidung eines Produktes aufgebaut werden kann, wird ebenfalls durch das Modell abgebildet: Hier sieht das Geschäftsmodell eine nachträgliche Gestaltung des Produktes voraus. Diverse Services und Produkteigenschaften sind nachträglich zubuchbar. Zum Kunden wird eine intensive Austauschbeziehung aufgebaut. Mit aktivem Marketing, durch Push-Benachrichtigungen auf dem Bordcomputer des Automobils, E-Mail-Benachrichtigungen, Zeitschriften oder Social Media ist diese intensive Art der Kundenbeziehung möglich. Kunden sind dadurch stets über aktuelle Angebot informiert. Die *beschriebene Informationsplattform ist für den Kunden zugleich die zentrale* Abwicklungsmöglichkeit zur nachträglichen Buchung. Über die zu erstellende Plattform werden Mikrotransaktionen sowie permanente Zubuchungen erledigt. Ob einmalige Buchung der Sitzheizung oder die permanente Freischaltung von mehr Reichweite innerhalb des Autos. Diese Optionen sind für den Kunden mit der Buchungsplattform gegeben. Im Zuge der großflächigen Verbreitung der Smartphones ist es auch hier für den Anwender von großer Bedeutung die Kompatibilität mit gängigen Betriebssystem herzustellen.

Der Baustein *Revenue Streams* beschreibt die Monetarisierung des Geschäftsmodells. In der Vergangenheit ist der Verkauf von Autos die dominierende Umsatzgenerierung gewesen. Durch innovative Bezahlmodelle, eigene Geschäftsbanken und weitere Ideen, konnte das Automobilgeschäft bisher große Erfolge verzeichnen (Berret 2016, S. 30). Die Zukunft ist jedoch ungewiss. Ein Grund liegt darin, dass nachfolgende Generationen und damit auch Zielgruppen dem Automobil als das Statussymbol einen immer geringeren Wert beimessen. Sinkende Absatzzahlen sind in den letzten Jahren lediglich durch die globale Expansion und den starken Absatzmarkt in China ausgeglichen worden. Der heimische deutsche Markt ist gesättigt. Auch diese Entwicklung ist ein Zeichen dafür, dass der Automobilmarkt innovative Konzepte fordert und benötigt (Stricker, Matthias, Tsang 2011, S. 6). Kunden erwarten individuell auf ihre Bedürfnisse zugeschnittene Lösungen, während die Automobilhersteller Lösungen für einen wachsenden Wettbewerb und damit stetig steigendem Kostendruck entwickeln müssen. Umsatzgenerierung alleine durch den Verkauf von Automobilen zu betreiben ist kein nachhaltiges Geschäftsmodell.

Der Siegeszug der Freemium-Geschäftsmodelle und deren In-App Käufe ist nicht ohne weiteres auf die Automobilindustrie extrapolierbar (Spencer 2016, abgerufen am 16.10.2017). Die grundlegende Idee dahinter und die damit vermittelte Freiheit für den Kunden ist die zentrale Innovation des Geschäftsmodells. Die eindimensionale Fokussierung auf den Verkauf von möglichst vielen Automobilen ist demnach kein zukunftsfähiges Konzept für die Automobilhersteller mehr. Services zur Individualisierung, On-Demand-Lösungen, Kundenfokus und komplette Individualität sowie Kontrolle für den Kunden bilden die Herausforderungen zur langfristigen Umsatzgenerierung (Rumpelt 2017, S. 9). Anstelle einer Einmalzahlung bei dem Verkauf eines Automobils ist es durch die softwaregesteuerte Individualisierung möglich, Umsatzkanäle dauerhaft mittels Immer-wieder-Zahlungen zu ermöglichen. Die Bereiche, die innerhalb des Automobils für eine solche On-Demand-Lösung in Frage kommen können, sind folgende:

- Beschleunigung, maximale Geschwindigkeit
- Reichweite des Fahrzeuges
- Antriebsart, Allradausstattung
- Infotainment, Navigation
- Soundsysteme, Smartphone Integration
- Konnektivität zu digitalen Diensten

Die *Key Resources* umfassen diverse Aspekte. Dieses Geschäftsmodell erfordert gewisse Voraussetzungen zur erfolgreichen Anwendung. Einer der wichtigsten Aspekte ist sicherlich der gedankliche Umbruch des klassischen Geschäftes. Blindverbau innerhalb der Automobilindustrie und nachträgliche Umsatzgenerierung stehen diametral zu den Quartalszielen vieler Unternehmen. Vertrauen in das System und die damit verbundenen Kapitalressourcen sind große Hürden, die es zu überwinden gilt. Auf der technologischen Seite ist vor allem die sogenannte OTA-Technologie eine Schlüsselkomponente (Frost & Sullivan 2016). OTA steht hierbei für Over-the-Air und bezeichnet die Aktualisierung und Installation von Software über das mit dem Internet verbundene Auto. Mit dieser ist die Nutzung des Angebots und die infrastrukturelle Ausstattung innerhalb des Autos eng verbunden. Kreativität und Entwicklungskompetenz ist zur Umsetzung eine Schlüsselressource, da durch die Kombination dieser beiden Attribute neue Lösungen entstehen. Eine weitere wichtige Ressource ist auch die Nutzung von Sharing-Angeboten für die eigenen Fahrzeuge. Kunden kommen mit den Fahrzeugen und dem neuen Geschäftsmodell in Kontakt, informieren sich über die Vorteile und lassen sich überzeugen.

Die *Key Activities* sind eng mit den *Key Resources* verbunden. Auf der technologischen Seite ist eine Etablierung der OTA-Technologie und den connected cars erfolgskritisch. Sharing services sind aufzubauen, die Servicestruktur muss etabliert werden und auch die enorm wichtige Plattform gilt es zu entwickeln. Außerdem ist die E-Mobilität ein großer Treiber dieses Geschäftsmodells. Elektrische Komponenten sind einfacher über Software zu steuern als mechanische. Die Motorleistung, Antriebsart etc. lassen sich somit einfacher als bei einem Verbrennungsmotor regeln. Zusätzlich proklamiere ich einen neuen Denkansatz für die Automobilindustrie. Losgröße 1, Toyota-Produktionssystem, Lean Management und weitere Denkansätze sind in der Vergangenheit beispiellos erfolgreich gewesen. Fragmentierte Supply Chains, globale Transportwege, Konsignationslager etc. sind das Resultat einer Komplexitätssteigerung innerhalb der Automobilindustrie über Jahre hinweg. Im Zuge der Digitalisierung und Anwendung dieses Geschäftsmodells wird eine Abkehr von diesen und ein Umdenken in Richtung „Design for Variants" gefordert. Mit Blindverbau, Big Data und dem Einsatz von Sprachsteuerung sowie intelligenten Lösungen zur Reduzierung von unnützen Bedienelementen, ist eine deutliche Komplexitätsreduktion möglich. Die Effekte die sich daraus ergeben sind in der ganzen Supply Chain zu verzeichnen (Wildemann 2008, S. 226). Eine schnellere Reaktion auf Marktentwicklungen, eine ausgeglichenere und agil agierende Produktion, die Nutzung von Skalierungseffekten, eine Anwendung von Pull-Systemen und die Reduktion von Lagerplatz und -kosten ist möglich.

Ohne die richtigen *Key Partners* ist das Konzept nicht umsetzbar. Die notwendigen Kooperationen umfassen die Zusammenarbeit des OEM mit Finanzinstituten für die Zahlungsabwicklung. Sie umfassen außerdem für die Gewährleistung der digitalen Infrastruktur, Sicherheit und Erreichbarkeit die Partnerschaft mit Cloud-Anbietern, Sicherheitsfirmen und der Regierung. Modullieferanten gewinnen an Verantwortung als Entwicklungspartner, Marketingfirmen haben erweiterten Spielraum für innovative Ansätze und Datenanalysefirmen sind für die richtige Nutzung der Daten verantwortlich. Ohne eigenen Sharing-Dienst ist auch die Kooperation mit einem Anbieter innerhalb dieses Geschäftsfeldes wichtig. Automobilhersteller sind normalerweise mit Kooperationen in den meisten genannten Gebieten ausgestattet und haben beste Voraussetzungen für die erfolgreiche Anwendung des Geschäftsmodells.

Neben den vorher vorgestellten Umsatzströmen ist es auch von großer Bedeutung, die Kostentreiber des Geschäftsmodells zu beleuchten. Die *Cost Structure* setzt sich aus unterschiedlichen Elementen zusammen. Auf der einen Seite sind der Aufbau, die Entwicklung und der Betrieb einer Plattform für die Zahlungsabwicklung, Buchung und den Service kostenintensiv. Der Aufbau eines eigenen Sharing-Services ist mit hohen Kosten verbunden genauso die Marketing-Aktionen, die zur Gewinnung und Begeisterung von Kunden notwendig sind. Das Geschäftsmodell fordert grundlegende Veränderungen innerhalb des Automobils. Diese zu entwickeln und umzusetzen erzeugt hohe Kosten für den Automobilhersteller. Langfristige Investitionen sind notwendig und auch der anfängliche Verlust, welcher sich einstellen kann, ist in die Betrachtung mit einzubeziehen. Die Sicherheit des Autos sowie die Datenauswertung gehören außerdem zu den wichtigsten Kostentreibern.

In der folgenden Abbildung 4 ist das erläuterte Business Modell Canvas einer softwaregesteuerten Individualisierung des Automobils zusammenfassend dargestellt.

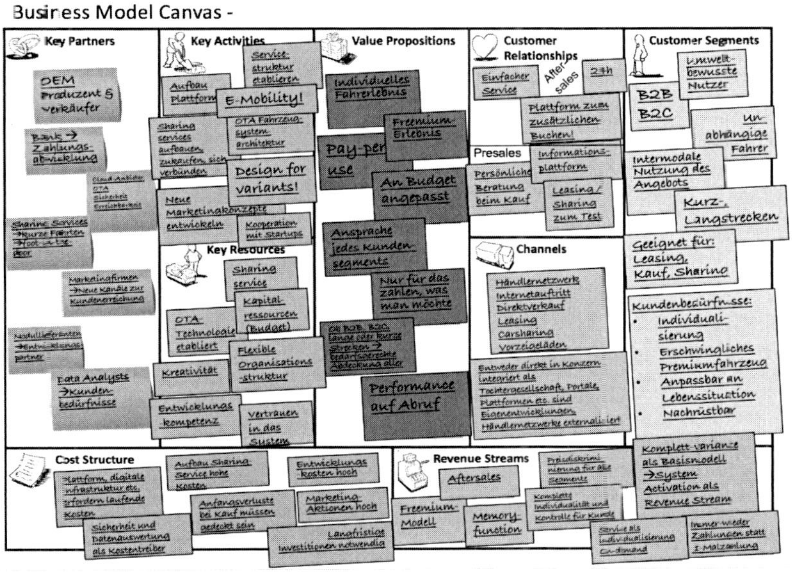

Abbildung 4: Das Business Model Canvas einer softwaregesteuerten Individualisierung des Automobils

3 Herausforderungen und Ausblick

Das hier aufgespannte Geschäftsmodell ist eine innovative Idee, um schwindende Gewinnmargen und gestiegene Komplexität in der Automobilindustrie zu reduzieren. Die Umstellung der klassischen Produktion auf Blindverbau, die Vernetzung aller Module und die Möglichkeit der nachträglichen Gestaltung für den Kunden sind Ansätze, die ein Umdenken in der Automobilindustrie fordern. Der Ansatz erscheint radikal, doch Tesla hat bereits mit der teilweisen Umsetzung eines ähnlichen Ansatzes großen Erfolg. Autopiloten, Batteriereichweite oder Motorleistung sind für einen teuren Aufpreis permanent nachträglich zubuchbar (Rumpel 2017, S. 9). Die Ausweitung dieses Angebotes für die Bereiche Infotainment, Komfortfunktionen, Sicherheit und weitere ist die notwendige Konsequenz aus einem sich verändernden Markt.

Die großen Themen der Digitalisierung spielen auch eine Rolle innerhalb dieses Geschäftsmodells. Datenschutz bei personalisierten Angeboten, Sharing-Dienstleistungen und den weiteren Komponenten des Geschäftsmodells sind Themen zahlreicher Diskussionen. Durch die Erweiterung des Autos als Lebensraum ist die Schnittstelle zu unserem Privatleben im Auto integriert. Die Integration des Smartphones bietet einen Angriffspunkt und auch bei der Nutzung digitaler Services im Internet ist keine 100%-tige Sicherheit gewährleistet. Außerdem ist die Sicherheit des Fahrzeuges bei einer kompletten Vernetzung ein kritischer Punkt.

Firewalls, abgetrennte Funktionalitäten, proaktives Software Management und weitere Maßnahmen sind von den Automobilherstellern anzuwenden. Kritische Funktionen wie Bremsen, Beschleunigung, autonome Fahrfunktionen können bei einer Kompromittierung durch Fremde großen Schaden anrichten.

Die Jurisdiktion sieht sich großen Veränderungen gegenüber. Neue Geschäftsmodelle verändern den Markt schneller, als die Rechtsprechung entscheiden kann. Die Anwendung von On-Demand-Funktionalitäten innerhalb des Autos stellt auch hier die Gesetzeslage vor neue Herausforderungen. Abrechnungen sind aus steuerlicher Sicht ein politisches Neuland, da entschieden werden muss, ob der mögliche Wert eines Autos, bei Aktivierung aller Module, oder der genutzte Wert eines Autos herangezogen wird. Auch das Thema Haftung spielt sich innerhalb dieser neuen Fragestellung ab. Im Schadensfall muss die Schadenshöhe bestimmt werden. Die Schadenshöhe kann sich hier auf den tatsächlich genutzten Wert des Autos beziehen oder auf den möglichen.

Die temporären Buchungsservices erfordern Forschungsbedarf in vielerlei Hinsicht. Lebenszyklen sind neu zu berechnen genauso der Bedarf und die Marktakzeptanz eines solchen Geschäftsmodells. A.T. Kearny geht in einer Untersuchung von Absatzsteigerungen von 40% bei Anwendung von On-Demand-Funktionalitäten aus sowie einem attraktiven Zusatzangebot für die Kunden (Rumpelt 2017, S. 9). Neue Preismodelle sind auf die neuen Marktanforderungen und Geschäftsgegebenheiten abzustimmen. Stündliche, tägliche oder andere Arten von Zubuchungen innerhalb des Autos gilt es geeignet zu Bepreisen. Die Preisdiskriminierung für die unterschiedlichen Kundensegmente und -bedürfnisse ermöglicht die Generierung von Umsätzen in ungeahnter Weise.

Das Beratungsunternehmen McKinsey geht in einem Ausblick für die Automobilindustrie von einem großen unerschlossenen neuen Markt aus. Mit 1,5 Billionen Euro ist der Markt mit vernetzten Services und neuen Geschäftsmodellen beziffert (McKinsey & Company 2016, S. 4). Genau in diesem Bereich findet das vorliegende Geschäftsmodell Anwendung.

Literatur

Becker, W., Ulrich, P. (2013): *Geschäftsmodelle im Mittelstand*. Stuttgart: Kohlhammer Verlag

Berret, M. (2016): *Transformation of the car industry*. München: Roland Berger GmbH - Automotive Competence Center. URL: https://www.rolandberger.com/publications/publication_pdf/roland_berger_acc_insights_1.pdf.

Bonnet, D., Buvat, J., Subrahmanyam, K.V.J. (2015): *When Digital Disruption Strikes - How Can Incumbents Respond?* In: Capgemini Consulting: Digital Transformation Review - Strategies for the Age of Digital Disruption. (S. 78-88). Capgemini Consulting.

Clarke, P., Mullancy, E. (2015): *What's next for global mobility? The impact of megatrends.*

PricewaterhouseCoopers GmbH. URL: https://www.pwc.com/gx/en/services/people-organisation/publications/assets/pwcmegatrends-and-mobility.pdf.

Diez, W. (2012): *Automobilwirtschaftliche Vertriebssysteme und die Rolle des Automobilhandels.*

In: Grundlagen der Automobilwirtschaft: das Standardwerk der Automobilbranche. (S. 275-307) München: Springer Automotive Media.

Diez, W. (2015): *Automobil-Marketing: erfolgreiche Strategien, praxisorientierte Konzepte, effektive Instrumente.* München: Vahlen.

Endlein, M. (2017): *DAT Report.* Ostfildern: Deutsche Automobil Treuhand GmbH.

Frost & Sullivan (2016): *Automotive Over-the-Air Updates.* Mountain View, USA: Frost & Sullivan.

Gassmann, O., Frankenberger, K., Csik, M. (2013): *Geschäftsmodelle entwickeln: 55 innovative Konzepte mit dem St. Galler Business Model Navigator.* München: Carl Hanser Verlag GmbH & Company KG.

Institut für Mobilitätsforschung (2010): *Zukunft der Mobilität Szenarien für das Jahr 2030.* München: Institut für Mobilitätsforschung. URL: https://www.ifmo.de/files/publications_content/2010/ifmo_2010_Zukunft_der_Mobilitaet_Szenarien_2030_de.pdf.

Kapteyn, C. (2012): *Optimierung des Geschäftsmodells.* Berlin: KAPTEYN Unternehmensberatung und Coaching.

Kerth, K., Asum, H., Stich, V. (2011): *Die besten Strategietools in der Praxis: Welche Werkzeuge brauche ich wann? Wie wende ich sie an? Wo liegen die Grenzen?* München: Hanser Verlag.

McKinsey & Company. (2016): *Automotive revolution - perspective towards 2030.*

Osterwalder, A. (2004): *The Business Model Ontology - A proposition in a design science approach.* Lausanne: Universite de Lausanne.

Osterwalder, A., Pigneur, Y., Wegberg J.T.A. (2011): *Business Model Generation: Ein Handbuch für Visionäre, Spielveränderer und Herausforderer.* Frankfurt am Main: Campus Verlag.

Porter, M.E. (1989): *How competitive forces shape strategy.* In: Asch, David und Cliff [eds.] Bowman (1989): *Readings in Strategic Management.* (S.133-144). Houndmills: Macmillan Education UK.

Ruff, F. (2012): *Die dreifache Neuerfindung des Automobils.* In: Innovationskulturen um das Automobil. Stuttgart: Mercedes-Benz Classic Archive, S. 209-224.

Rumpelt, T. (2017): *Schöne neue Kaufwelten: Autokauf 4.0: Digital Pricind und On-Demand-Geschäftsmodelle.* In: A.T. Kearny: Automotive (S. 6-11). Düsseldorf: A.T.

Kearny

Rusnjak, A, (2014): *Entrepreneurial Business Modeling: Definitionen - Vorgehensmodell - Framework - Werkzeuge - Perspektiven*. Wiesbaden: Springer Gabler.

Schade, W. et al (2012): *Zukunft der Automobilindustrie*. Berlin: Büro für Technikfolgen-Abschätzung beim deutschen Bundestag. URL: https://www.tabbeim-bundestag.de

Schallmo, D. (2012): *Geschäftsmodell-Innovation: Grundlagen, bestehende Ansätze, methodisches Vorgehen und B2B-Geschäftsmodelle*. Berlin, Heidelberg: Springer Fachmedien

Schallmo, D. (2013): *Geschäftsmodelle erfolgreiche entwickeln und implementieren*. Berlin, Heidelberg: Springer Gabler. isbn: 978-3-642-37993-2.

Schawel, C., Billing, F. (2014): *Top 100 Management Tools: das wichtigste Buch eines Managers; von ABC-Analyse bis Zielvereinbarung*. Wiesbaden: Springer Gabler.

Spencer, G. (2016): *Exploring the App Store's Top Grossing Chart*. URL: https://www.macstories.net/stories/exploring-the-app-stores-top-grossingchart/ (besucht am 16. 10. 2017).

Stricker, K., Matthias, G., Tsang, R. (2011): *Vom Automobilbauer zum Mobilitätsdienstleister*. Germany / Switzerland: Bain & Company.

Stüber, J. (2017): *Extras auf Zeit - wie die Autoindustrie ein neues Bezahlmodell ausheckt*. Berlin. URL: http://ngin-mobility.com/artikel/servicedemand-auto/ (besucht am 05. 12. 2017).

Weinreich, U. (2016): *Lean digitization: digitale Transformation durch agiles Management*. Berlin; Heidelberg: Springer Gabler.

Wildemann, H. (2008): *Variantenmanagement: Leitfaden zur Komplexitätsreduzierung, -beherrschung und -vermeidung in Produkt und Prozess*. München: TCW Transfer-Centrum GmbH & Co. KG

Winterhoff, M. (2009): *Zukunft der Mobilität 2020 - Die Automobilindustrie im Umbruch?* Arthur D. Little. URL: www.adl.com/mobilitaet-2020-lang.

Wirtz, B. (2013): *Business Model Management: Design - Instrumente - Erfolgsfaktoren von Geschäftsmodellen*. Wiesbaden u.a.: Gabler Verlag.

Journal of Engineering, Management and Operations 1 (2018)

Industrie 4.0 und Recht – ein Überblick

Joachim Gschwinder[a]

[a] *ESB Business School, Reutlingen University, D-72762 Reutlingen*

Abstract

Der Beitrag gibt einen Überblick über die Zusammenhänge zwischen Industrie 4.0 und Recht. Ausgangspunkt sind die vertragsrechtlichen Grundlagen mit den Fragen, ob und wie zwischen Maschinen überhaupt Verträge zustande kommen können, wie Industrie 4.0-Verträge vertragstypologisch einzuordnen sind und welche Grenzen bei der Verwendung Allgemeiner Geschäftsbedingungen bestehen können. Anschließend werden die Themen dargestellt, die derzeit die Diskussion um das Haftungsrecht in der Industrie 4.0 beherrschen. Im Rahmen des Datenrechts wird geklärt, wem Daten zustehen und wie der Datenschutz geregelt ist. Zudem wird aufgezeigt, welche rechtlichen Vorgaben zur Gewährleistung von IT-Sicherheit im Unternehmen bestehen. Schließlich werden auch arbeitsrechtliche Herausforderungen von Industrie 4.0 angesprochen, wie die zunehmende „Entbetrieblichung", das Aufkommen neuer Beschäftigungsformen, die Forderung nach mehr Flexibilisierung der Arbeitszeit sowie Fragen des Mitbestimmungsrechts.

1 Einleitung

Industrie 4.0 ist derzeit in aller Munde. Nach der Erfindung der Dampfmaschine, der Einführung der Fließbandproduktion und dem Siegeszug des Computers hat die vierte industrielle Revolution („Industrie 4.0") bereits begonnen. Diese ist gekennzeichnet durch die Verzahnung der Produktion mit modernster Informations- und Kommunikationstechnik sowie durch die digitale Vernetzung der cyber-physischen Systeme (CPS) zu einer „intelligenten Fabrik" („Smart Factory"). Menschen, Maschinen und Produkte kommunizieren in Echtzeit direkt miteinander, so dass daraus „intelligente Wertschöpfungsketten" entstehen. Industrie 4.0 jedoch nur als rein technische Herausforderung zu begreifen, wäre viel zu verkürzt. Neben die Fragen der technischen Machbarkeit treten eine Vielzahl von weiteren Themenfeldern. Diese betreffen etwa die gesellschaftlichen, politischen, wirtschaftlichen, aber auch die ethischen Fragen, die sich im Zusammenhang mit einer digitalisierten Produktion stellen. Auch rechtliche Aspekte spielen bei Industrie 4.0 eine große Rolle. Ohne Kenntnis und Beachtung der einschlägigen Rechtsvorschriften wird Industrie 4.0 nicht funktionieren. Eine rechtzeitige Auseinandersetzung auch mit den rechtlichen Themen ist daher unabdingbar für jeden, der Industrie 4.0 im Unternehmen einführen und anwenden möchte. Wie sich nicht nur beim autonomen Fahren, sondern auch bei zahlreichen Anwendungen der Industrie 4.0 zeigt, ist die technische Entwicklung bereits weiter fortgeschritten als das Recht. Das Recht „hinkt hinterher". Zahlreiche rechtliche Fragen sind derzeit noch unbeantwortet und bedürfen einer Klärung.

Das Ziel dieses Beitrags ist, einen ersten Überblick über die wesentlichen Rechtsthemen rund um die Industrie 4.0 zu geben. Entsprechend der Konzeption des Sammelbands soll insbesondere auch bei Nicht-Juristen ein Bewusstsein für diese Themen und deren Bedeutung geschaffen werden. Hinzuweisen ist darauf, dass die Auswahl der behandelten Rechtsgebiete (Vertragsrecht, Haftungsrecht, Datenrecht, IT-Sicherheit und Arbeitsrecht) keinen abschließenden Charakter hat. Zwar handelt es sich dabei um die Themen, welche aktuell die Diskussion um das Verhältnis von Recht und Industrie 4.0 im Wesentlichen dominieren. Es soll aber nicht unerwähnt bleiben, dass es darüber hinaus noch zahlreiche weitere für Industrie 4.0 bedeutsame Rechtsthemen gibt (siehe Sassenberg & Faber 2017), auf die in diesem Beitrag nicht eingegangen werden kann.

2 Vertragsrecht

2.1 Ausgangslage

Verträgen kommt im Kontext von Industrie 4.0 eine besondere Bedeutung zu. Zunächst soll geklärt werden, welche Konsequenzen entstehen, wenn selbststeuernde Maschinen miteinander „kommunizieren". Insbesondere wird der Frage nachgegangen, ob Maschinen auch rechtsverbindliche Verträge abschließen können. Die mit Industrie 4.0 einhergehende Zunahme individualisierter Verträge und die Kombination von Verträgen mit zusätzlichen Internetservices lässt außerdem die Frage aufkommen, wie solche Verträge rechtlich einzuordnen sind. Für zahlreiche Rechtsfragen der digitalisierten und vernetzten Produktion gibt es derzeit noch keine gesetzlichen Regelungen. Unternehmen sind deshalb verstärkt darauf angewiesen, Rechtssicherheit mittels vertraglicher Regelungen zu erreichen. Die Vertragsfreiheit als Element der Privatautonomie kann dabei jedoch an die Grenzen des Rechts der Allgemeinen Geschäftsbedingungen stoßen.

2.2 Vertragsschluss zwischen Maschinen (M2M-Verträge)

Typisch für Industrie 4.0 ist die automatisierte Steuerung von Produktionsabläufen durch eine Vernetzung von Maschinen und IT-Systemen. So sind Maschinen bereits heute in der Lage, autonom miteinander zu kommunizieren. Beispielsweise ist es möglich, dass eine automatische Fertigungsanlage im Unternehmen A während des Fertigungsprozesses autonom einen Bestellvorgang für ein Ersatzteil im Unternehmen B auslöst. In diesem Zusammenhang muss geklärt werden, ob bei einem solchen

Bestellvorgang Willenserklärungen abgegeben werden, die zu einem wirksamen Vertragsschluss führen, wem diese Willenserklärungen gegebenenfalls zuzurechnen sind und ob eine fehlerhafte Willenserklärung durch Anfechtung wieder beseitigt werden kann.

Das deutsche Zivilrecht kennt keine speziellen gesetzlichen Regelungen für die Kommunikation zwischen Maschinen. Nach den Regelungen des Bürgerlichen Gesetzbuchs (BGB) setzt ein Vertrag (mindestens) zwei übereinstimmende Willenserklärungen (Angebot und Annahme) voraus. Eine Willenserklärung ist jedoch immer eine menschliche Willensäußerung, die auf die Herbeiführung eines rechtlichen Erfolgs gerichtet ist. Da Maschinen nicht über eine eigene Rechtspersönlichkeit verfügen, können sie nicht Träger von Rechten und Pflichten sein und sind deshalb auch nicht in der Lage, individualisierte Willensäußerungen von sich zu geben.

Daher sind maschinengesteuerte Erklärungen demjenigen zuzurechnen, aus dessen Sphäre die Erklärung stammt und der sich bei Vertragsschluss eines autonomen Systems bedient bzw. ein autonomes System in die Vertragsdurchführung einbringt. Das heißt, grundsätzlich werden dem Betreiber der Maschine die Inhalte einer Maschinenerklärung zugerechnet. Rechtstechnisch werden IT-Systeme, die automatisierte Erklärungen abgeben, als Boten behandelt, die lediglich die Willenserklärung ihres Betreibers überbringen (Hornung & Hofmann 2017a, S. 202). Der Maschine kommt dabei nur die Funktion einer (technischen) Empfangs- oder Erklärungsvorrichtung zu. Mithin lässt sich feststellen, dass sich der „Vertragsschluss zwischen Maschinen" mit den Regeln des BGB zufriedenstellend erklären lässt und derzeit noch keine speziellen gesetzlichen Regelungen für Maschinenerklärungen benötigt werden (ausführlich dazu Heuer-James, Chibanguza & Stücker 2018, S. 2820 ff.). Hilfreich wäre allenfalls eine gesetzliche Klarstellung, wonach die Vorschriften des BGB über Willenserklärungen auch dann gelten, wenn sie unter Verwendung von Maschinen erfolgen (Plattform Industrie 4.0 2016, S. 7). Da die Kommunikation innerhalb selbstlernender und selbstoptimierender Maschinen und Systeme für den Betreiber jedoch immer weniger vorhersehbar wird, darf indes bezweifelt werden, ob die Zurechnung an den Maschinenbetreiber auch in Zukunft Bestand haben wird. Mit der fortschreitenden technischen Entwicklung wird dann möglicherweise eine neue rechtliche Bewertung von Maschinenerklärungen erforderlich sein (Bräutigam & Klindt 2015, S. 1138). Es ist auch zu erwarten, dass vertragliche Regelungen zunehmend an Bedeutung gewinnen, wonach Maschinenreaktionen ein rechtlicher Gehalt beigemessen wird oder der Empfänger einer Maschinenerklärung den Zugang einer Nachricht zu bestätigen hat (Bitkom 2016a, S. 35).

Die Inhalte einer Maschinenerklärung sind nach derzeitigem Rechtsverständnis für den Betreiber einer Maschine auch dann verbindlich, wenn er diese Inhalte nicht voraussehen konnte. Eine Ausnahme soll nur für den Fall gelten, dass die fehlerhaften Erklärungsinhalte für den Empfänger klar erkennbar waren. Fehlerhafte Maschinenerklärungen können demnach nur nach den allgemeinen Regeln des BGB und bei Vorliegen der entsprechenden (Anfechtungs-)Voraussetzungen beseitigt werden. Grundsätzlich muss sich der Betreiber auch eine fehlerhafte Erklärungstätigkeit der Maschine zurechnen lassen. Die Anfechtungsvoraussetzungen sind nur gegeben, wenn fehlerhafte Angaben des Betreibers in die Maschinenerklärung übernommen wurden.

Hornung & Hofmann (2017a, S. 203) beschreiben diese Problematik anhand eines anschaulichen Beispiels: „Bestellt also eine automatisierte Fertigungsanlage zu viel Material, weil das intelligente Warenlager aufgrund eines technischen Fehlers einen zu geringen Stand gemeldet hat, ist dies kein Grund zur Anfechtung. Das gleiche gilt, wenn der Bedarf falsch berechnet wurde. Gibt der Betreiber den Inhalt der Erklärung jedoch vor und definiert nur die Bedingungen, unter der sie abgegeben wird (wenn Menge unter x fällt, dann 100 Teile nachbestellen) und wird der Inhalt dann falsch übermittelt (1000 Teile bestellen), berechtigt ihn dies, sich durch Anfechtung wegen falscher Übermittlung nach § 120 BGB von der Willenserklärung wieder zu lösen (...)." Sofern die Anfechtungsvoraussetzungen nicht vorliegen, kommen bei verbindlichen, nicht vorhersehbaren („fehlerhaften") Erklärungsinhalten derzeit nur Schadensersatzansprüche gegen Dritte (den Maschinenhersteller) in Betracht. Denkbar wäre auch in diesem Zusammenhang, mögliche „Fehlerfälle" im Voraus zwischen Hersteller und Betreiber vertraglich zu regeln und noch weitere Vertragsaufhebungsgründe vorzusehen. Dabei sind allerdings die Restriktionen durch das Recht der Allgemeinen Geschäftsbedingungen (§§ 305 ff. BGB) (vgl. 2.4) zu beachten.

2.3 Vertragsgegenstand

Die stetig zunehmende Möglichkeit der Individualisierung von Produkten durch Industrie 4.0 (bis hin zu Losgröße 1-Fertigungen) stellt ein Abrücken von der klassischen Massenproduktion dar. Das Kaufrecht droht damit an Bedeutung zu verlieren. Verträge über die Lieferung von Einzelanfertigungen wären somit nicht mehr als Kaufvertrag (§§ 433 ff. BGB), sondern als Werklieferungsvertrag (§ 650 BGB) zu qualifizieren. Zwar findet auf den Werklieferungsvertrag ebenfalls das allgemeine Kaufrecht Anwendung, jedoch sieht § 650 BGB mehrere Modifikationen vor. Ob und inwieweit in diesem Zusammenhang ein gesetzgeberisches Eingreifen erforderlich sein wird, bleibt abzuwarten.

Bei gemischttypischen Verträgen, also Kaufverträgen, die mit Internetservices kombiniert sind („Smart Things"), kommt es für die vertragstypologische Einordnung auf den Leistungsschwerpunkt an. Wenn kein Leistungsschwerpunkt feststellbar ist, sind die Einzelleistungen entscheidend. Sofern der Gegenstand nur in Kombination mit einem Internetservice (Dienst) genutzt werden kann, handelt es sich um einen reinen Kaufvertrag, bei dem der Dienst vertraglicher Bestandteil ist. Ein Abschalten des Dienstes würde die ganze Sache mangelhaft machen. Da in diesem Fall der Mangel jedoch nicht bereits bei Gefahrübergang (§ 434 Abs. 1 BGB) vorliegt, erscheint fraglich, ob überhaupt auf das kaufrechtliche Sachmängelrecht zurückgegriffen werden kann (Bräutigam & Klindt 2015, S. 1138). Ist der Gegenstand auch ohne den (Internet-)Dienst nutzbar, so handelt es sich bei dem Erwerb des Gegenstands um einen Kaufvertrag, die Dienste unterliegen dann aber einem Dienstvertrag. Sofern in diesem Fall der Internetdienst eingestellt wird, muss – mangels spezieller Gewährleistungsvorschriften im Dienstvertragsrecht – eine rechtliche Beurteilung nach dem allgemeinen Leistungsstörungsrecht (§§ 280 ff. BGB) erfolgen.

2.4 Vertragsfreiheit und Allgemeine Geschäftsbedingungen

Um innovative Industrie 4.0-Geschäftsprozesse rechtlich zu gestalten und zu steuern, bedarf es zunehmend vertraglicher Regelungen zwischen den beteiligten Unternehmen. Dabei kommt einer effizienten Rechtsgestaltung durch (vorformulierte) Standardverträge eine besondere Bedeutung zu. Standardisierte Verträge geraten jedoch auch im B2B-Verkehr zwischen Unternehmen zunehmend in Konflikt mit den Einschränkungen des Rechts der Allgemeinen Geschäftsbedingungen (AGB). Die für Verbrauchergeschäfte im B2C-Verkehr vorgesehenen Klauselverbote (§§ 308, 309 BGB) werden von der Rechtsprechung über die Generalklausel des § 307 BGB zunehmend auch im unternehmerischen Geschäftsverkehr angewendet. Die Grundannahme des AGB-Rechts, Verträge zwischen Unternehmen nur einer eingeschränkten AGB-Kontrolle zu unterziehen (vgl. § 310 Abs. 1 BGB), findet in der Rechtsprechung immer weniger Beachtung. Die ausufernde Anwendung von Vorschriften, die eigentlich dem Verbraucherschutz dienen, auch auf Verträge zwischen Unternehmen verhindert zunehmend belastbare vertragliche Regelungen und entpuppt sich dabei als Standortnachteil. Dies kann insbesondere im internationalen Wirtschaftsverkehr dazu führen, dass Vertragspartner durch die Wahl einer ausländischen Rechtsordnung der ausufernden deutschen AGB-Rechtsprechung zu entgehen versuchen („Flucht aus dem deutschen Recht"). Diese Vorgehensweise ist jedoch nicht geeignet, das Vertrauen in den Standort Deutschland und dessen Rechtsordnung zu stärken. Der Grundsatz der Vertragsfreiheit sollte daher dringend wieder stärker berücksichtigt werden. Dies könnte etwa durch eine gesetzliche Anerkennung von Industriestandardverträgen oder durch eine ergänzende Klarstellung in § 310 Abs. 1 BGB erfolgen (Bitkom 2016b, S. 14).

3 Haftungsrecht

3.1 Haftungssysteme und Schadensszenarien

Die zunehmende Automatisierung der Produktion sowie die Vernetzung von Wertschöpfungssystemen rufen eine Reihe haftungsrechtlicher Fragen hervor. Insbesondere die verbundmäßige Interaktion von autonomen cyber-physischen Systemen (CPS) lässt zu den bekannten Schadensszenarien neue hinzutreten. Denkbar sind etwa Störungen der Produktion, des Betriebsablaufs oder der Supply Chain, beispielsweise durch Verbindungsunterbrechungen in der M2M-Kommunikation, als Folge von Fehlfunktionen in „Smart Products" oder durch das Weitergeben fehlerhafter Daten. Auch die fehlerhafte Erhebung und Analyse von Big Data können gravierende Auswirkungen haben und zu Schäden innerhalb der Wertschöpfungskette führen.

Das deutsche Haftungsrecht unterscheidet zwischen vertraglicher Haftung und außervertraglicher (deliktischer) Haftung. Vertragsrechtliche Schadensersatzansprüche aus § 280 BGB kommen insbesondere dann in Betracht, wenn der Nutzer bzw. Betreiber zugleich auch der Käufer eines Produkts ist. Der Anspruch aus § 280 Abs. 1 BGB setzt ein wirksames Schuldverhältnis (z.B. einen Vertrag) sowie eine Pflichtverletzung des Anspruchsgegners voraus. Aber auch wenn keine vertraglichen Beziehungen bestehen, sind Schadensersatzansprüche möglich. Die außervertragliche Haftung greift etwa dann, wenn zwischen Hersteller und Nutzer (Betreiber) eines Produkts eine weitere Person (z.B. Händler) zwischengeschaltet ist und der Nutzer (Betreiber) nur zu dieser Person eine vertragliche Beziehung hat. Darüber hinaus trifft das deutsche Haftungsrecht eine Unterscheidung zwischen der Verschuldenshaftung (verschuldensabhängige Haftung) und der Haftung ohne Verschulden (verschuldensunabhängige Haftung). Vertragliche und außervertragliche (deliktische) Haftungsansprüche setzen in der Regel ein Verschulden des Anspruchsgegners voraus. Dieses Verschulden muss grundsätzlich der Geschädigte beweisen (beim vertraglichen Anspruch aus § 280 Abs. 1 wird gemäß Satz 2 dieser Vorschrift das Verschulden des Anspruchsgegners vermutet). Nur ausnahmsweise kommt eine verschuldensunabhängige Gefährdungshaftung in Betracht.

Für Sach- und Personenschäden, die durch ein fehlerhaftes Produkt verursacht werden, sieht das deutsche Recht Ansprüche aus der deliktsrechtlichen Produzentenhaftung (§ 823 Abs. 1 BGB) und aus dem Produkthaftungsgesetz (§ 1 ProdhaftG) vor. Ein bedeutsamer Unterschied zwischen diesen beiden Anspruchsgrundlagen liegt in der Voraussetzung des Verschuldens (bei der deliktischen Produzentenhaftung nach § 823 Abs. 1 BGB ist ein Verschulden erforderlich, die Produkthaftung nach dem Produkthaftungsgesetz ist verschuldensunabhängig).

3.2 Verschulden in der Smart Factory

Das zentrale Thema im Bereich der Haftung ist somit das „Verschulden“. Verschulden heißt Verantwortlichkeit für vorsätzliches oder fahrlässiges Verhalten (§ 276 BGB). Insbesondere bei autonom handelnden Systemen ist aber unklar, wer für etwaige Schäden aufgrund welcher rechtlichen Grundlage einzustehen hat. Wenn beispielsweise eine Maschine im Sinne einer „predictive maintenance“ autonom Wartungsbedürftigkeit meldet und sich dies als unzutreffend erweist, kann dies zu Produktionsausfällen und weiteren Störungen innerhalb der Supply Chain führen. Wer muss dafür einstehen?

Wenn Maschinen miteinander kommunizieren und Prozesse automatisiert ablaufen, kann es schwierig sein, natürliche Personen als Auslöser oder Urheber einer Handlung festzustellen. Mangels eigener Rechtspersönlichkeit kommt eine Haftung von Maschinen allerdings nicht in Betracht. Im deutschen Haftungsrecht kann ein Schaden nur einem Menschen, nicht aber einer Maschine zugerechnet werden. Das menschliche Verhalten muss für den konkreten Schaden kausal gewesen sein. In der rechtswissenschaftlichen Literatur wird zwar zum Teil die Ansicht vertreten, es müsse – etwa im Bereich der Robotik – eine Art elektronische Person mit begrenzter Rechtssubjektivität und eigenen Verantwortlichkeiten geschaffen werden (z.B. Beck 2009, S. 229 f.). Diese Ansicht stößt jedoch auf Widerspruch. So weist Spindler (2016, S. 816) darauf hin, dass auch eine „maschinelle Person“ von einem Menschen als „Organ“ vertreten werden müsse. Auch Hornung und Hofmann (2017b, S. 13) sehen in der Einführung einer elektronischen Person kaum lösbare Probleme, da eine Maschine (im Gegensatz zum Menschen) keinen Anreiz habe, ihre Haftungsmasse nicht willkürlich zu gefährden. Ein eigenständiges Haftungssubjekt in Form einer elektronischen Person ist tatsächlich nicht erforderlich, da die bestehenden Haftungskonzepte ausreichende Lösungen für die unterschiedlichen Schadensszenarien bieten (so auch Bundesverband der Deutschen Industrie e.V.& Noerr 2015, S. 14). Da somit nur menschliches Handeln als Anknüpfungspunkt für eine Haftung in Betracht kommt, ist beim Haftungsvorwurf grundsätzlich auf das (schuldhafte) Inverkehrbringen der Maschine oder die Entscheidung zur tatsächlichen Nutzung abzustellen. Auf eine spätere autonome, den Schaden letztlich auslösende „Entscheidung" der Maschine kommt es dabei nicht an.

Entscheidend ist und bleibt dabei die Vorhersehbarkeit der schädigenden Kausalität aus Sicht des Herstellers bzw. des späteren Verwenders (Bitkom 2016a, S. 39). Wenn das erlernte Verhalten der Software nicht vorhersehbar ist, wird man diesen meist kein Verschulden vorwerfen können. Bei vollständig autonomen Systemen, bei denen sich der Verwender in den Datenaustausch von Abläufen nicht mehr einschalten kann und ihm keine Entscheidungshoheit bzw. Eingriffsmöglichkeit mehr bleibt (z.B. bei selbstlernenden Industrierobotern), gerät das bestehende Haftungssystem an seine Grenzen (Bundesverband der Deutschen Industrie e.V.& Noerr 2015, S. 14). Eine Verantwortungszuordnung zumindest nach den bekannten Kausalitäts- und Zurechnungsprinzipien ist dann nicht mehr möglich, so dass fraglich ist, ob in diesen Konstellationen noch an der Verschuldenshaftung festgehalten werden kann. Eine in der Literatur vorgeschlagene Lösung dieses Problems besteht in der Einführung einer Gefährdungshaftung (z.B. Bräutigam & Klindt 2015, S. 1138 f.). Gefährdungshaftungshaftungstatbestände sieht das Gesetz für Fälle vor, in denen mit einer erlaubten Tätigkeit unvermeidlich und typischerweise eine gewisse Gefährdung einhergeht. Die Gefährdungshaftung beruht auf der Erwägung, dass derjenige, der eine gefährliche Betätigung ausübt oder eine gefährliche Anlage betreibt, für den Schaden einstehen soll, der bei einer Realisierung der Gefährdung entsteht (z.B. Haftung des Fahrzeughalters beim Betrieb eines Kraftfahrzeugs gemäß § 7 Straßenverkehrsgesetz oder die Haftung des Tierhalters nach § 833 Bürgerliches Gesetzbuch). Es obliegt jedoch allein dem Gesetzgeber, für den Einsatz künstlicher Intelligenz einen neuen Gefährdungshaftungstatbestand zu schaffen. Bräutigam & Klindt (2015, S. 1139) schlagen darüber hinaus auch für den Bereich der Big Data-Analyse einen verschuldensunabhängigen Gefährdungstatbestand vor, um so die dort ebenfalls anzutreffenden Schwierigkeiten bei der Bestimmung des Verschuldens und der Kausalität zu beseitigen. Die weitere Entwicklung in diesen Bereichen sollte aufmerksam verfolgt und gegebenenfalls durch entsprechende gesetzgeberische Maßnahmen geregelt werden.

3.3 Haftungszurechnung

Aufgrund der Komplexität der interagierenden Systeme kann es im Rahmen von Industrie 4.0-Anwendungen problematisch sein festzustellen, wer bzw. welches System letztlich den Schaden verursacht hat. Wenn in einem solchen Schadensfall verschiedene Beteiligte involviert sind, kann die Klärung der Verursachungs- und Verschuldensfrage schwierig oder auch unmöglich werden. Lässt sich nicht ermitteln, welcher von mehreren Beteiligten mit seinem Beitrag den Schaden verursacht hat, empfiehlt es sich im Deliktsrecht, die gesamtschuldnerische Haftung aller in Betracht kommenden Verursacher nach § 830 Abs. 1 S. 2 BGB heranzuziehen (so z.B. Bitkom 2016a, S. 39; Hornung & Hofmann 2017b, S. 14).

Im Rahmen der arbeitsteiligen Produktionsorganisation einer Smart Factory kann zudem vertraglich festgelegt werden, wie die Verteilung der Haftungsbeiträge intern ausgestaltet werden soll. Mit solchen vertraglichen Regelungen kann – vor allem auch im Bereich der vertraglichen Haftung – das schwer vorhersehbare Zusammenwirken komplexer Systeme geregelt werden. Zu beachten ist, dass auch hier die getroffenen vertraglichen Regelungen den Grenzen des AGB-Rechts unterliegen. Wesentliche Vertragspflichten können demgemäß nur aus schwerwiegenden Gründen abbedungen werden. Insbesondere ist es nicht möglich, die Haftung bei typischerweise vorhersehbaren Schäden – etwa für entgangenen Gewinn oder Nutzungsausfall – auf grobe Fahrlässigkeit zu beschränken oder komplett auszuschließen (Hornung & Hofmann 2017a, S. 201).

4 Datenrecht

4.1 Abgrenzung

Daten werden heute als „Rohstoffe des 21. Jahrhunderts" bezeichnet und stellen als solche einen beachtlichen wirtschaftlichen Wert dar (Paal & Hennemann 2017, S. 1697 mit weiteren Nachweisen). Sie sind die Grundlage für die Entwicklung und Verbesserung vorhandener Produkte, Dienstleistungen und Geschäftsmodelle sowie für Effizienzsteigerungen in Produktionsprozessen. Damit sind Daten auch entscheidende Faktoren für die Wettbewerbsfähigkeit von Unternehmen. Die Datenvielfalt bei Industrie 4.0 ist enorm; es können Datensätze über eine Maschine bestehen, genauso wie Daten, die bei der Nutzung einer Maschine anfallen (z.B. Daten über die Konfigurierung von Maschinen, Nutzungsdaten der Maschine, Umgebungsdaten, Daten mit Bezug zu den Arbeitnehmern, die an der Maschine arbeiten, usw.). Ein (Daten-)Mehrwert kann zudem dadurch erzielt werden, dass durch die Korrelation und Auswertung von riesigen Datensätzen neue Erkenntnisse für die Produktion generiert („Big Data-Analyse") und dadurch Produktionsabläufe weiter optimiert werden.

Im Datenrecht ist zwischen Daten mit Bezug zu einer konkreten natürlichen Person („personenbezogene Daten") und Daten ohne Bezug zu einer konkreten natürlichen Person („Maschinendaten") zu unterscheiden. Nur personenbezogene Daten unterliegen dem Datenschutzrecht. Handelt es sich um Maschinendaten (wie z.B. Informationen über die Betriebsstunden, den Druck, die Temperatur oder den Stromverbrauch einer Maschine), ist das Datenschutzrecht nicht einschlägig. Insbesondere bei den für die Wertschöpfung besonders bedeutsamen Maschinendaten ist von großer Bedeutung, wem diese Daten zuzuordnen sind und wer daraus Rechte ableiten kann.

4.2 Datenhoheit

Wenn eine Maschine während der Produktion Daten generiert, stellt sich die Frage, wem diese Daten zustehen: Dem Hersteller, der die mit einem Datenerfassungsgerät versehene Maschine liefert oder dem Betreiber, der die Daten durch Nutzung der Maschine generiert? Konflikte über die Datenhoheit können auch bei der Durchführung von Big Data-Analysen durch einen externen Dienstleister entstehen. Ungeklärt ist auch hier, welche Eigentumsrechte an den Daten begründet werden und ob der Auftraggeber gegenüber dem externen Dienstleister bezüglich der aufbereiteten Daten einen Herausgabeanspruch hat.

§ 903 BGB gibt dem Eigentümer einer Sache das Recht, mit der Sache nach Belieben zu verfahren (Nutzungsrecht) und andere von jeder Einwirkung auszuschließen (Abwehrrecht). Sachen sind gemäß § 90 BGB aber nur „körperliche Gegenstände". Da Daten diese Körperlichkeit gerade fehlt, fallen sie nicht unter den zivilrechtlichen Eigentumsbegriff.

Ausführlich wird in der rechtswissenschaftlichen Literatur (vgl. etwa die Nachweise bei v. Baum, Appt & Schenk 2017a, S. 1826 f.) deshalb diskutiert, wie Daten dennoch einem bestimmten Rechtsträger zugewiesen werden können. Unter anderem wird – abgeleitet aus der strafrechtlichen Verfügungsbefugnis des § 303a Strafgesetzbuch – in Analogie zu § 903 BGB ein „Dateneigentum" konstruiert. Dieses soll demjenigen zustehen, der den Skripturakt über die Speicherung der Daten vornimmt (Hoeren 2013, S. 490). Teilweise wird die Ansicht vertreten, über das „Recht des Datenbankherstellers" gemäß §§ 87a ff. Urheberrechtsgesetz (UrhG) eine Zuordnung der Daten vornehmen zu können (Hornung & Hofmann 2017a, S. 194 f.). Diskutiert wird auch, ob über den in den §§ 17 und 18 UWG (Gesetz gegen den unlauteren Wettbewerb) geregelten Schutz von Geschäfts- und Betriebsgeheimnissen eine Berechtigung an den Daten hergeleitet werden kann. Diese Ansätze vermögen jedoch nicht zu überzeugen. Mangels einer planwidrigen Regelungslücke im Immaterialgüterrecht verbleibt kein Raum für Analogien (Sattler 2017, S. 43, Rn. 47). Auch die Heranziehung des Urheberrechtsgesetzes führt nicht weiter. Die §§ 87a ff. UrhG schützen nicht die einzelnen Daten, sondern nur die „Datenbank". Eine Datenbank gemäß § 87a Abs. 1 S. 1 UrhG ist eine Sammlung von Werken, Daten oder anderen unabhängigen Elementen, die systematisch oder methodisch angeordnet und einzeln mit Hilfe elektronischer Mittel oder auf andere Weise zugänglich sind und deren Beschaffung, Überprüfung oder Darstellung eine nach Art oder Umfang wesentliche Investition erfordert. Diese Voraussetzung dürfte nur in einigen Fällen erfüllt sein. Gerade bei Big Data ist dies zweifelhaft, zumal es sich dabei eher um einen „Datenhaufen" als um eine systematische oder methodische Anordnung handelt. Zudem stünde das Recht an der Datenbank auch nur demjenigen zu, der das Investitionsrisiko in die Herstellung der Datenbank trägt. Schließlich schützen auch die §§ 17, 18 UWG nur den Berechtigten von Geschäfts- und Betriebsgeheimnissen, begründen selbst aber keine Berechtigung.

Die Diskussion ist weiter im Fluss. Es ist festzuhalten, dass die beschriebenen (und auch die weiteren in der Literatur diskutierten) Ansätze weder ein Eigentumsrecht an Daten noch ein dem Eigentum vergleichbares Recht mit absoluter Wirkung begründen. Gesetzliche Regelungen über die Zuweisung der Datenhoheit bestehen derzeit nicht. Es muss daher weiter den Unternehmen überlassen bleiben, mittels vertraglicher Vereinbarungen die Nutzungsrechte an Daten zu regeln. Der Gesetzgeber sollte hier zurückhaltend sein und nicht übereilte gesetzliche Regelungen erlassen (so auch Bundesverband der Deutschen Industrie e.V. & Noerr 2015, S. 12; Plattform Industrie 4.0 2016, S. 22). Sollte dies in der Praxis zu Verwerfungen führen, kann der Gesetzgeber immer noch eingreifen und nachbessern.

4.3 (Beschäftigten-)Datenschutz

4.3.1 Ausgangssituation

Industrie 4.0 und die sich daraus ergebenden immensen Anwendungsmöglichkeiten führen zu einer nahezu unendlichen Flut an Daten. Im Rahmen von „Big Data" sollen mit Hilfe von Analyse-Algorithmen aus riesigen Datenmengen neue Erkenntnisse gewonnen werden, die es ermöglichen, bisher unbekannte Zusammenhänge darzulegen und Prognosen für bestimmte Vorgänge zu erstellen (Niklas & Thurn 2017, S. 1589). Zu denken ist auch an die vielfachen Einsatzmöglichkeiten von so genannten „Wearables" – wie etwa Datenbrillen zur Kommissionierung von Waren. Auch so genannte Assistenzsysteme, die Vitaldaten von Beschäftigten auslesen und damit die Durchlaufgeschwindigkeit einer Fertigungsstraße steuern können, sind heute technisch längst machbar. Dies kann zum einen zu Steigerungen der Produktivität oder zur Verbesserung der Arbeitssicherheit führen. Zum anderen eignen sich diese „Wearables" in der Regel auch zur Kontrolle der Beschäftigten (Weichert 2017, S. 565 ff.). Sofern personenbezogene Daten verarbeitet werden, ist der Anwendungsbereich des Datenschutzrechts eröffnet. Im Folgenden soll der Fokus auf den wichtigen Bereich des Beschäftigtendatenschutzes gerichtet werden. Hier stehen sich regelmäßig die widerstreitenden Interessen des Arbeitgebers und des Arbeitnehmers gegenüber. Der Arbeitgeber möchte möglichst viele Informationen bzw. Daten über den Arbeitnehmer haben. Der Arbeitnehmer hat ein berechtigtes Interesse, dass seine Daten nur insoweit erhoben und verarbeitet werden, als dies tatsächlich notwendig und ein ausreichender Schutz der persönlichen Daten gewährleistet ist.

Das Datenschutzrecht ist geprägt von einem Ineinandergreifen des europäischen und des deutschen Rechts. Seit dem 25.05.2018 gilt in der Europäischen Union die EU-Datenschutz-Grundverordnung (DSGVO). Die Zulässigkeit der Verarbeitung personenbezogener Daten im Rahmen von Industrie 4.0 muss sich daran messen lassen. Als EU-Verordnung ist die DSGVO in den Mitgliedsstaaten unmittelbar anwendbares Recht und bedarf keiner Umsetzung in nationales Recht. Zwar enthält die DSGVO keine spezifischen Regelungen zum Beschäftigtendatenschutz. Die allgemeinen Rechtfertigungsgründe (Art. 6 DSGVO), die Auskunftspflichten (Art. 12 ff. DSGVO) und die Pflicht zur Datenlöschung (Art. 17 DSGVO) sind jedoch auch auf den Beschäftigtendatenschutz anwendbar (Neighbour 2017, S. 235, Rn. 114). Von der in der DSGVO mittels zahlreicher Öffnungsklauseln vorgesehenen Möglichkeit, den Beschäftigtendatenschutz zu spezifizieren (Art. 88 Abs. 1 DSGVO), hat der deutsche Gesetzgeber Gebrauch gemacht und ein neues Bundesdatenschutzgesetz (BDSG) verabschiedet. Das BDSG ergänzt, konkretisiert und modifiziert die DSGVO; es ist ebenfalls am

25.05.2018 in Kraft getreten.

4.3.2 Personenbezug

Die Regelungen des Datenschutzes gelten nur für die Verarbeitung personenbezogener Daten. Personenbezogene Daten sind nach § 46 Nr. 1 BDSG „alle Informationen, die sich auf eine identifizierte oder identifizierbare natürliche Person (betroffene Person) beziehen; als identifizierbar wird eine natürliche Person angesehen, die direkt oder indirekt (...) identifziert werden kann" (vgl. auch Art. 4 Nr. 1 DSGVO). Von einer Identifizierbarkeit ist dann auszugehen, wenn die Daten zuverlässig auf eine konkrete Person schließen lassen (Redeker 2017, Rn. 932). Ein Personenbezug kann aber auch indirekt hergestellt werden, so zum Beispiel, wenn aus Maschinendaten Informationen über natürliche Personen bzw. deren Verhalten abgeleitet werden können (Mantz & Spittka 2017, S. 147 f., Rn. 22). Bei sämtlichen Mensch-Maschine-Interaktionen, wie sie zahlreiche Industrie 4.0-Anwendungen beinhalten, dürften alle dabei generierten Daten den jeweiligen Mitarbeitern zuordenbar sein, z.B. durch den Abgleich mit Schichtplänen oder weil sich die Mitarbeiter zuvor an der Maschine einloggen müssen. Aber auch schon die Verknüpfung von Sensordaten mit anderen Datenquellen können zur Bestimmbarkeit bzw. Identifizierbarkeit einer Person führen (v. Baum, Appt & Schenk 2017a, S. 1825 f.). Ein indirekter Personenbezug kann aber auch durch die Verknüpfung von selbst nicht personenbezogenen Daten hergestellt werden. Aus diesem Grund sind große, für Big Data-Analysen vorgesehene Datenbestände latent immer von einem Personenbezug „bedroht" (Hornung & Hofmann 2017b, S. 17).

Sobald Daten einen (direkten oder indirekten) Personenbezug aufweisen, ist für die Nutzung, Verwendung für andere Zwecke oder die Weitergabe der Daten an Dritte eine gesetzliche Ermächtigungsrundlage oder die vorherige Einwilligung der Betroffenen erforderlich.

4.3.3 Zulässigkeit der Datenverarbeitung

Die Verarbeitung personenbezogener Daten ist nach Art 6 DSGVO nur zulässig, wenn ein gesetzlicher Erlaubnistatbestand vorliegt oder die betroffene Person eingewilligt hat. Das Bundesdatenschutzgesetz weicht von diesen Prinzipien nicht grundlegend ab. Gemäß § 26 Abs. 1 S. 1 BDSG ist die Verarbeitung personenbezogener Daten von Beschäftigten gestattet, wenn sie für die Entscheidung über die Begründung eines Beschäftigungsverhältnisses, dessen Durchführung oder Beendigung oder zur Aufhebung oder Erfüllung der sich aus einem Gesetz oder einer Betriebsvereinbarung ergebenden Rechte und Pflichten des Betriebsrats erforderlich ist. Auch zur Aufklärung von Straftaten dürfen Daten verarbeitet werden (§ 26 Abs. 1 S. 2 BDSG). Ebenso können Betriebsvereinbarungen als Rechtfertigungsgrund für die Verarbeitung von Beschäftigtendaten herangezogen werden (§ 26 Abs. 4 BDSG), müssen sich aber an den Anforderungen des Art. 88 Abs. 2 DSGVO orientieren. Es ist dabei immer darauf zu achten, dass die Datenverarbeitung dem Verhältnismäßigkeitsprinzip sowie den Grundsätzen der Zweckbindung, der Datenvermeidung und der Datensparsamkeit entspricht. Sofern eine Verarbeitung personenbezogener Daten nicht verhältnismäßig ist, lässt sich diese auch nicht mittels einer Betriebsvereinbarung rechtfertigen. Vor diesem Hintergrund ist eine Rechtfertigung von umfangreichen Big-Data-Analysen mittels einer Betriebsvereinbarung als eher kritisch zu beurteilen (Niklas & Thurn 2017, S. 1594).

Sofern kein gesetzlicher Rechtfertigungsgrund vorliegt, ist für die Verarbeitung von personenbezogenen Daten die vorherige Einwilligung des Betroffenen zwingende Voraussetzung (Art. 6 Abs. 1 lit. a DSGVO und § 26 Abs. 2 BDSG). Die Einwilligung muss auf einer freien Entscheidung des Betroffenen beruhen. Insbesondere muss der Betroffene seine Entscheidung frei von jeglichem Zwang treffen können. Dies kann zweifelhaft sein, wenn dem Betroffenen die Einwilligung unter Ausnutzung einer Machtposition abgenötigt wurde. Dabei muss in einem Arbeitsverhältnis der Arbeitgeber die bestehende Abhängigkeit des Beschäftigten berücksichtigen. Entscheidend ist letztlich, unter welchen Umständen der Arbeitnehmer die Einwilligung erteilt. Von einer Freiwilligkeit der Einwilligung ist gemäß § 26 Abs. 2 BDSG grundsätzlich dann auszugehen, wenn für den Arbeitnehmer ein rechtlicher oder wirtschaftlicher Vorteil erzielt wird bzw. wenn Arbeitgeber und Arbeitnehmer gleichgelagerte Interessen haben (dazu Wybitul 2017, S. 416).

Eine Einwilligung ist zudem nur dann möglich, wenn der Betroffene den Umfang der vorgesehenen Datenverarbeitung kennt. Der Betroffene muss daher über den Zweck der Datenverarbeitung vorab informiert werden (Grundsatz der informierten Einwilligung; § 51 Abs. 4 S. 3 BDSG). Es darf bezweifelt werden, ob dieses weite Einwilligungserfordernis bei Big Data-Analysen vorliegen kann. Insbesondere müsste dann auch der Analysezweck vorab bekannt gemacht werden, was sich aber bei Big Data-Analysen aufgrund fehlender Bestimmtheit als schwierig erweist (Hieke 2017, S. 19). Eine weitere Unsicherheit im Zusammenhang mit der Einwilligung resultiert aus der Möglichkeit, dass diese – auch wenn sie freiwillig erteilt wurde – jederzeit vom betroffenen Arbeitnehmer widerrufen werden kann (vgl. § 51 Abs. 3 BDSG; Art. 7 Abs. 3 DSGVO). In der Konsequenz bedeutet dies, dass eine Einwilligung als Grundlage der Datenverarbeitung nur in Ausnahmefällen in Betracht kommen wird (Schmidl & Tannen 2017, S. 1638).

Ein möglicher Weg, dem Anwendungsbereich des Datenschutzrechts zu entgehen, liegt in der Anonymisierung der Daten. Dabei müssen personenbezogene Daten dergestalt verändert werden, dass eine Zuordnung nicht mehr oder nur mit unverhältnismäßig großem Aufwand möglich ist (v. Baum, Appt & Schenk 2017a, S. 1826). Neuen Aufschwung erleben die Ansätze von Datenschutz durch Technikgestaltung („Privacy by Design“) und Datenschutz durch datenschutzfreundliche Voreinstellungen („Privacy by Default“) mit deren Aufnahme in Art. 25 Abs. 1 und 2 DSGVO.

5 IT-Sicherheitsrecht

5.1 Ausgangslage

Cyber-Attacken durch Hacker und Cyber-Kriminelle stellen für die vernetzte Fabrik ein besonderes Risiko dar. Der unberechtigte Zugriff auf IT-Systeme (z.B. durch Einschleusen von Malware oder Abgriff von Daten) kann zu enormen Schäden führen. Durch Kaskadeneffekte innerhalb der vernetzten Produktion können dadurch die internen Abläufe gestört werden. Mantz und Spittka (2017, S. 174, Rn. 139) schildern eindrücklich mögliche Szenarien: „So finden Logistikroboter oder –drohnen möglicherweise ihren Weg nicht und liefern Produkte nicht rechtzeitig oder falsch aus. Zusammenstöße oder Abstürze können zu Ausfällen und (sogar Personen-)Schäden führen. Ferner können Geräte, die sich selbst um ihre Wartung kümmern, indem sie rechtzeitig Ersatzteile bestellen oder entsprechenden Bedarf melden, so manipuliert werden, dass sie die Meldung oder Bestellung ‚vergessen' und mangels Wartung dann für längere Zeit ausfallen. Auf der anderen Seite könnten durch einen Angreifer Wartungsintervalle unnötig verkürzt werden, was ebenfalls häufigere und damit längere Ausfallzeiten und insbesondere höhere Kosten nach sich zieht." Die Sicherstellung von IT-Sicherheit gehört damit zu den Kernthemen der Digitalisierung. Dabei handelt es sich in erster Linie um eine technische Aufgabe, die aber von entsprechenden rechtlichen Rahmenbedingungen flankiert sein muss.

Generell haben der Vorstand einer Aktiengesellschaft und der Geschäftsführer einer Gesellschaft mit beschränkter Haftung die Pflicht, für eine ordnungsgemäße Geschäftsorganisation zu sorgen (vgl. § 91 Abs. 2 Aktiengesetz und § 43 Abs. 1 GmbH-Gesetz), wozu auch notwendige Schutzvorkehrungen zur Gewährleistung von IT-Sicherheit gehören. Klar definierte IT-Sicherheitsstandards lassen sich daraus jedoch nicht ableiten. Außerdem haben § 91 Abs. 2 Aktiengesetz und § 43 Abs. 1 GmbH-Gesetz keine drittschützende Wirkung. Bei einem Pflichtenverstoß des Vorstands bzw. des Geschäftsführers können zwar der Gesellschaft selbst Schadensersatzansprüche zustehen, nicht aber Dritten, wie beispielweise Kunden oder Zulieferern (v. Baum, Appt & Schenk 2017b, S. 1888 f.).

Die Gesetzeslage zur IT-Sicherheit ist insgesamt unübersichtlich. Sie ist gekennzeichnet durch „eine (fragmentarische) Gemengelage aus mehreren unterschiedlichen Rechtsmaterien" (Bräutigam & Klindt 2015, S. 1141). Für Systeme, die personenbezogene Daten verarbeiten, sehen § 64 BDSG sowie Art. 32 DSGVO datenschutzorientierte Maßnahmen zur Gewährleistung von IT-Sicherheit vor. Wer personenbezogene Daten verarbeitet, muss technische und organisatorische Maßnahmen ergreifen, um die Einhaltung dieser Datenschutzgesetze zu gewährleisten.

Mit dem IT-Sicherheitsgesetz (ITSiG) schuf der deutsche Gesetzgeber im Jahr 2015 zahlreiche Regelungen zur Verbesserung der IT-Sicherheit, die nicht auf die Verarbeitung personenbezogener Daten beschränkt sind. Ein Jahr später erließ der europäische Gesetzgeber die Richtlinie EU/2016/1148 zur Gewährleistung einer hohen Netzwerk- und Informationssicherheit (NIS-Richtlinie). Darin werden Maßnahmen zur Gewährleistung eines hohen gemeinsamen Sicherheitsniveaus von Netz- und Informationssystemen innerhalb der Europäischen Union festgelegt. Die Umsetzung dieser Richtlinie in das nationale Recht machte eine Anpassung des ITSiG notwendig.

5.2 Gesetz über das Bundesamt für Sicherheit in der Informationstechnik (BSIG)

Die Umsetzung der Richtlinie EU/2016/1148 und die Anpassung des ITSiG führte zu Änderungen in verschiedenen Einzelgesetzen, wie z.B. im Gesetz über das Bundesamt für Sicherheit in der Informationstechnik (BSIG). Darin sind die Sicherheitsanforderungen und Meldepflichten für Betreiber so genannter „Kritischer Infrastrukturen" definiert.

Der Begriff der „Kritischen Infrastruktur" ist in § 2 Abs. 10 BSIG legal definiert. Dazu gehören die im Gesetz aufgeführten „Sektoren" (Energie, Informationstechnik und Telekommunikation, Transport und Verkehr, Gesundheit, Wasser, Ernährung sowie Finanz- und Versicherungswesen). Die Relevanz dieses Gesetzes für Industrie 4.0 - Anwendungen hängt somit zum einen davon ab, ob einer der genannten Sektoren betroffen ist. Zum anderen kommt es darauf an, ob die – in einer besonderen Rechtsverordnung (BSI-Kritisverordnung) aufgelisteten – branchenspezifischen Schwellenwerte erreicht werden (Hornung & Hofmann 2017a, S. 207). Alle vom Anwendungsbereich erfassten Betreiber müssen die genannten IT-Sicherheitsmaßnahmen nach dem „Stand der Technik" umsetzen. Nach § 8a Abs. 1 BSIG sind sie dazu verpflichtet, „angemessene organisatorische und technische Vorkehrungen zur Vermeidung von Störungen der Verfügbarkeit, Integrität, Authentizität und Vertraulichkeit ihrer informationstechnischen Systeme, Komponenten oder Prozesse zu treffen, die für die Funktionsfähigkeit maßgeblich sind". Alle zwei Jahre ist für die betroffenen Betreiber darüber hinaus ein verpflichtendes Sicherheitsaudit vorgesehen (§ 8a Abs. 3 BSIG). Zudem greifen bei „erheblichen" Störungen der IT-Systeme bestimmte Meldepflichten an das Bundesamt für Sicherheit in der Informationstechnik (vgl. § 8b Abs. 4 BSIG). Sofern eine vernetzte Fabrik nicht zu den im Gesetz aufgezählten Sektoren gehört – oder der Schwellenwert nicht erreicht ist -, finden die Vorschriften keine unmittelbare Anwendung. Denkbar ist jedoch, dass die in den Gesetzen beschriebenen Standards vertraglich weitergegeben werden (v. Baum, Appt & Schenk 2017b, S. 1889).

6 Arbeitsrecht

6.1 Ausgangslage

Durch die fortschreitende Digitalisierung und Vernetzung werden Arbeitsformen, Arbeitsstrukturen und Arbeitsinhalte in den Unternehmen grundlegend verändert. Industrie 4.0 ist ohne Arbeit 4.0 nicht denkbar. Mit der Veröffentlichung des „Grünbuchs Arbeiten 4.0" (Bundesministerium für Arbeit und Soziales, 2015) stieß das Bundesministerium für Arbeit und Soziales im Jahr 2015 den Dialog um das Thema „Arbeiten 4.0" an. Damit wurde eine breite Diskussion in Gang gesetzt. Der im November 2016 vom Bundesministerium für Arbeit und Soziales veröffentlichte Diskussionsentwurf eines „Weißbuchs Arbeiten 4.0" (Bundesministerium für Arbeit und Soziales,

2017) fasst verschiedene Gestaltungsvorschläge zusammen, die seit mehreren Jahren die Diskussion um dieses Thema beherrschen, wie etwa den Umbau der Arbeitslosenversicherung zur Arbeitsversicherung, die Einführung eines Wahlarbeitszeitgesetzes, die Ausweitung der gesetzlichen Rentenversicherung auf Selbständige und vieles mehr.

Auch in der Rechtswissenschaft beschäftigt man sich seit einigen Jahren mit dieser Thematik, Tendenz steigend. Auf dem 71. Deutschen Juristentag in Essen (2016) widmete sich die Abteilung Arbeits- und Sozialrecht ausschließlich der Digitalisierung der Arbeitswelt (Krause, 2016). In der rechtswissenschaftlichen Diskussion wird insbesondere auch der Frage nachgegangen, ob „Arbeit 4.0" auch ein „Arbeitsrecht 4.0" benötigt oder ob die Herausforderungen von Industrie 4.0 mit dem bestehenden Arbeitsrecht bewältigt werden können. Die zentralen Aspekte sind hierbei der Wandel der betrieblichen Strukturen und die damit einhergehende Entgrenzung von Arbeitsbeziehungen, der Umgang mit neuen Beschäftigungsformen, die Flexibilisierung der Arbeitszeit sowie Arbeitnehmerdatenschutz und betriebliche Mitbestimmung.

6.2 Wandel der betrieblichen Strukturen

Unter einem „Betrieb" versteht man im Arbeitsrecht eine organisatorische Einheit, innerhalb derer ein Arbeitgeber allein oder mit seinen Arbeitnehmern bestimmte arbeitstechnische Zwecke fortgesetzt verfolgt (Küttner/Kreitner 2017, S. 786).

Dieser schon klassische Betriebsbegriff wird im Rahmen der Digitalisierung jedoch mehr und mehr ausgehöhlt. Die Betriebsstrukturen der Arbeitswelt 4.0 mit ihrer Vernetzung verschiedener Unternehmen entfernen sich zunehmend davon (Kalbfus 2017, S. 11). Vor allem global handelnde Unternehmen machen von modernen Kommunikationsmöglichkeiten Gebrauch und bilden zunehmend unternehmensübergreifend funktionale Einheiten für einzelne Fachbereiche (wie z.B. HR, Finance, Vertrieb oder IT) mit eigenen Zuständigkeiten und Verantwortlichkeiten. Dezentrale spezialisierte und flexible Einheiten treten damit an die Stelle des Betriebs; es entstehen Matrixstrukturen, die zunehmend auch die klassischen Konzernstrukturen verdrängen (Günther & Böglmüller 2015, S. 1026; Kalbfus 2017, S. 11).

Diese Matrixstrukturen sind dadurch gekennzeichnet, dass die fachliche und die disziplinarische Führung der Mitarbeiter auseinanderfallen. Der Arbeitgeber, mit dem der Arbeitnehmer den Arbeitsvertrag geschlossen hat, hat das disziplinarische Weisungsrecht. Das fachliche Weisungsrecht übt hingegen die Führungskraft eines anderen Konzernunternehmens aus. Die „Entbetrieblichung" (Hanau 2016, S. 2614) führt also dazu, dass das Weisungsrecht nicht mehr vom Arbeitgeber, sondern zunehmend von unternehmensfremden Personen ausgeübt wird. Solche Matrixstrukturen, in denen mitbestimmungspflichtige Entscheidungen von einem anderen Unternehmen als dem Arbeitgeber-Unternehmen getroffen werden, stellen das Betriebsverfassungsrecht vor große Herausforderungen.

Ein „gemeinsamer Betrieb" im Sinne von § 1 Abs. 2 Betriebsverfassungsgesetz (BetrVG) ist in diesen Fällen nicht anzunehmen. Zwar kann innerhalb der Matrixorganisation regelmäßig von einem einheitlichen arbeitstechnischen Zweck ausgegangen werden. Es fehlt dagegen die für einen gemeinschaftlichen Betrieb erforderliche einheitliche Leitungsmacht bezüglich personeller und sozialer Angelegenheiten (Neighbour 2017, S. 219 Rn. 32).

Für Matrixstrukturen könnte eine Lösung in der Schaffung alternativer Organisationsstrukturen bestehen, wie sie in § 3 Abs. 1 BetrVG vorgesehen sind (Meyerhans 2017, S. 27 ff.). Dem steht jedoch entgegen, dass das Erfordernis eines Tarifvertrags solche alternativen betriebsverfassungsrechtlichen Organisationsstrukturen schwerfällig und unflexibel macht (Günther & Böglmüller 2015, S. 1027 f.). Hornung & Hofmann (2017b, S. 18) regen an, entsprechend den in § 3 Abs. 1 Nr. 4 BetrVG genannten Arbeitsgemeinschaften die Gründung einer zusätzlichen Arbeitnehmervertretung – mit Beteiligungsrechten – zu ermöglichen.

6.3 Neue Beschäftigungsformen

Neben das „Normalarbeitsverhältnis" treten mit dem Fortschreiten der Digitalisierung zunehmend „atypische Beschäftigungsformen" wie etwa das „Crowdworking". Dabei schreiben Unternehmen einzelne Projekte oder Arbeitsaufgaben über webbasierte Plattformen an eine unbekannte Masse (crowd) von Marktteilnehmern aus. Für Unternehmen bietet Crowdworking die Möglichkeit, schnell und flexibel auf schwankende Auftragseingänge reagieren zu können. Zudem kann dadurch ein Wissens- und Talentpotential – etwa im kreativen Bereich – genutzt werden, das möglicherweise in der eigenen Belegschaft nicht vorhanden ist. Für die Crowdworker besteht indessen ein Risiko im Fehlen sozialer Absicherung (Meyer-Michaelis, Falter & Schäfer 2016, S. 2543 f.).

Zu unterscheiden ist zwischen internem und externem Crowdworking. Beim externen Crowdworking handelt es sich um öffentlich zugängliche Plattformen, bei denen sich grundsätzlich jede Person um Aufträge bemühen kann. Beim internen Crowdworking nutzt das Unternehmen oder der Konzern eine interne Plattform und spricht damit eigene Arbeitnehmer an (Köhler 2017, S. 63).

Mangels Arbeitnehmereigenschaft scheidet eine Anwendung des Arbeitsrechts auf (externe) Crowdworker aus. Die Arbeitsaufgabe der Crowdworker ist regelmäßig in der entsprechenden Ausschreibung abschließend definiert; arbeitnehmertypischen Weisungen ist der Crowdworker somit gerade nicht ausgesetzt. Sofern der Crowdworker von einem einzigen Auftraggeber wirtschaftlich abhängig und einem Arbeitnehmer vergleichbar sozial schutzbedürftig ist, kann ein Crowdworker auch nach bereits geltendem Recht eine so genannte „arbeitnehmerähnliche Person" sein, mit der Folge, dass bestimmte arbeitsrechtliche Vorschriften anwendbar wären. Dies dürfte jedoch die Ausnahme darstellen, da Crowdworker in der Regel auf mehreren Crowdworking-Plattformen aktiv sind und für mehrere Auftraggeber tätig werden. Auch eine Einordnung als Heimarbeiter nach dem Heimarbeitsgesetz kommt wegen der regelmäßig fehlenden Abhängigkeit von einem Auftraggeber nicht in Betracht (Köhler 2017, S. 70 f.).

(Externe) Crowdworker sind somit (Solo-)Selbständige, deren Rechtsverhältnis nicht nach Arbeitsrecht, sondern nach Dienst- oder Werkvertragsrecht zu beurteilen ist. Dies bedeutet, dass Arbeitnehmerschutzgesetze wie etwa das Arbeitszeitgesetz, das Kündigungsschutzgesetz, das Entgeltfortzahlungsgesetz, das Bundesurlaubsgesetz, aber auch das Mindestlohngesetz keine Anwendung finden.

Zum Teil wird daher über eine Reform des Beschäftigtenschutzes und eine Neujustierung des Arbeitnehmerbegriffs durch Einführung eines allgemeinen Beschäftigtenbegriffs nachgedacht (Hanau 2016, S. 2616) Auch von Seiten der Gewerkschaften wird der Ruf nach einem Sonderschutz für Crowdworker lauter (Köhler 2017, S. 77 f.). Andere sehen indes keinen Handlungsbedarf und lehnen einen Sonderschutz als „unsystematische Forderung" ab (Günther & Böglmüller 2015, S. 1030).

6.4 Flexibilisierung der Arbeitszeit

Das öffentlich-rechtliche Arbeitszeitgesetz (ArbZG) dient dem Arbeitsschutz und bezweckt die Sicherheit und den Gesundheitsschutz von Arbeitnehmern, indem es die höchstzulässige Arbeitszeit begrenzt sowie zu Ruhepausen und Ruhezeiten verpflichtet. Dabei ist eine werktägliche Höchstarbeitszeit von acht Stunden vorgesehen. Diese kann auf zehn Stunden verlängert werden, wenn innerhalb von sechs Kalendermonaten oder innerhalb von 24 Wochen im Durchschnitt acht Stunden werktäglich nicht überschritten werden (§ 3 ArbZG). Zudem sieht das Gesetz zwingend vorgeschriebene Ruhepausen während der Arbeit (§ 4 ArbZG), eine ununterbrochene Ruhezeit von mindestens elf Stunden nach Beendigung der täglichen Arbeitszeit (§ 5 Abs. 1 ArbZG) sowie ein nur wenige Ausnahmen zulassendes Verbot von Sonn- und Feiertagsarbeit (§§ 9 ff. ArbZG) vor.

Die Nutzung moderner Kommunikationsmittel lässt die Grenzen zwischen Arbeit und Privatleben immer weiter verschwimmen. Internetfähige Smartphones, Tablet-PCs und Laptops machen es möglich, zeit- und ortsunabhängig für das Unternehmen Arbeitsleistungen zu erbringen. Während dies für Arbeitgeber immense Möglichkeiten der Flexibilisierung eröffnet, kann diese "Entgrenzung" bei Arbeitnehmern zu (gesundheitlichen) Belastungen führen. Gleichzeitig eröffnet mehr selbstbestimmtes Arbeiten Arbeitnehmern aber auch die Chance, Familie, Arbeit und Freizeit im Sinne einer „Work-Life-Balance" besser aufeinander abzustimmen. Auch das Bundesministerium für Arbeit und Soziales (2017, S. 73 ff.) geht in seinem „Weißbuch" davon aus, dass das Bedürfnis nach zeit- und ortssouveränem Arbeiten auf Seiten der Beschäftigten zunehmen wird.

Das starre Arbeitszeitgesetz ist jedoch kaum in Einklang zu bringen mit den Bedürfnissen nach einer flexiblen Lebensgestaltung. Die im Arbeitszeitgesetz geregelten Höchstarbeitszeiten und Ruhezeiten entsprechen heute nicht mehr den tatsächlichen Anforderungen. So lässt eine nur geringfügige Unterbrechung der Ruhezeit durch ein kurzes Telefonat oder die Antwort auf eine E-Mail die Mindestruhezeit von elf Stunden nach § 5 ArbZG von Neuem beginnen. Hier müssen die sich im Rahmen der Digitalisierung verändernden Bedürfnisse der Unternehmen und der Beschäftigten austariert werden, um so zum einen Flexibilität durch Nutzung moderner Medien zum Wohle der Unternehmen und der Beschäftigten zuzulassen und zum anderen den Risiken, die in der „Entgrenzung" von Arbeit und Privatleben liegen, vorzubeugen.

Das aktuelle Arbeitsrecht stellt dafür nur wenige Instrumente zur Verfügung. Die so genannte „Arbeit auf Abruf" (§ 12 Teilzeit- und Befristungsgesetz) führt hier nicht weiter, da der Einsatz vom Arbeitgeber mindestens vier Tage im Voraus angekündigt werden muss und sich damit für kurzfristige Arbeitseinsätze im Rahmen von Industrie 4.0 gerade nicht eignet (Neighbour 2017, S. 223, Rn. 50). Eher geeignet ist das Modell der so genannten „Vertrauensarbeitszeit“. Dieses zeichnet sich dadurch aus, dass der Arbeitgeber auf die Kontrolle der täglichen Arbeitszeit komplett verzichtet; für den Arbeitgeber ist allein das Erbringen der Arbeitsergebnisse maßgeblich (Küttner-Poeche 2017, S. 470 f.). Dennoch ist der Arbeitgeber auch bei der Vertrauensarbeitszeit verpflichtet, die Einhaltung der Ruhezeiten zu gewährleisten. Zudem treffen den Arbeitgeber Aufzeichnungspflichten hinsichtlich der über die werktägliche Arbeitszeit von acht Stunden hinausgehenden Arbeitszeit (§ 16 Abs. 2 ArbZG).

In der arbeitsrechtlichen Literatur gibt es mittlerweile zahlreiche Vorschläge, wie das enge Korsett des Arbeitszeitrechts im Sinne einer Flexibilisierung gelockert werden könnte. So wird unter Hinweis auf die europäische Arbeitszeitrichtlinie (2003/88/EG) vorgeschlagen, die tägliche Höchstarbeitszeit auf eine wöchentliche Höchstarbeitszeit umzustellen (Günther & Böglmüller 2015, S. 1028; Hanau 2016, S. 2617). Eine nur geringfügige Unterbrechung der Ruhezeit sei unbeachtlich (Hanau 2016, a.a.O.; so auch Krause 2016, B 46). Auch das restriktive Verbot der Sonn- und Feiertagsarbeit wird in Frage gestellt (Günther & Böglmüller 2015, S. 1028; Neighbour 2017, S. 222 f., Rn 48; Plattform Industrie 4.0 2016, S. 24).

Das Bundesministerium für Arbeit und Soziales (2017, S. 124) stellt indes im „Weißbuch“ klar, dass „eine allgemeine Öffnung des Arbeitszeitgesetzes wie eine Abkehr von der Norm des 8-Stunden-Tages zugunsten nur noch einer Wochenhöchstarbeitszeit (...) mit den Zielen des Arbeitsschutzes und der Zeitsouveränität nicht vereinbar [ist]“. Vorgeschlagen wird hingegen ein „Wahlarbeitszeitgesetz“ mit mehr Wahloptionen für Beschäftigte bei Arbeitszeit und Arbeitsort mit einer „konditionierten Möglichkeit der Abweichung von den geltenden Regelungen des Arbeitszeitgesetzes“.

Vor diesem Hintergrund bleibt spannend, ob und wann der Gesetzgeber den richtigen Zeitpunkt für ein Eingreifen sieht und welche Änderungen zur Flexibilisierung der Arbeitszeit tatsächlich vorgenommen werden. Handlungsbedarf besteht allemal.

6.5 Mitbestimmung

Im Rahmen von Industrie 4.0 stehen den Betriebsräten weitreichende Informations-, Beteiligungs- und Mitbestimmungsrechte zu. Von erheblicher Bedeutung ist dabei das Mitbestimmungsrecht gemäß § 87 Abs. 1 Nr. 6 Betriebsverfassungsgesetz (BetrVG) bei der Einführung und Anwendung technischer Einrichtungen, die dazu bestimmt sind, das Verhalten oder die Leistung der Arbeitnehmer zu überwachen.

Dabei genügt es nach der Rechtsprechung des Bundesarbeitsgerichts (ständige Rechtsprechung seit BAGE 27, 256), dass die technische Einrichtung objektiv geeignet ist, das Leistungsverhalten der Arbeitnehmer zu überwachen. Davon wird in Zeiten von Industrie 4.0 nahezu jeglicher Einsatz von Hard- und Software erfasst, auch wenn diese nicht zur Überwachung der Mitarbeiter eingesetzt werden (Giese 2017, S. 132). In der Praxis führt die weite Auslegung des § 87 Abs. 1 Nr. 6 BetrVG dazu, dass für fast alle Arten von IT-Tools eine Betriebsvereinbarung (oder eine Regelungsabrede) getroffen wird.

In der arbeitsrechtlichen Literatur wird zum Teil von einem „uferlosen" Mitbestimmungsrecht gesprochen, das den Entwicklungen der Industrie 4.0 entgegenstehe (Giese 2017, S.132; Neighbour 2017, S. 229, Rn. 82). Es wird deshalb vorgeschlagen, sich mehr am Wortlaut des § 87 Abs. 1 Nr. 6 BetrVG zu orientieren („bestimmt") und nicht schon die Datenerzeugung, sondern erst deren Auswertung der Mitbestimmung zu unterwerfen (Giese 2017, S. 139; Günther & Böglmüller 2015, S. 1027; Hanau 2016, S. 2615; Neighbour 2017, S. 229, Rn. 82). Krause (2016, B 80) plädiert hingegen für eine Ausweitung des Mitbestimmungsrechts auf alle Formen des Umgangs mit Daten, die sich Beschäftigten zuordnen lassen. Dies würde im Ergebnis zu einem Vetorecht des Betriebsrats gegenüber jeder Form von Digitalisierung im Arbeitsverhältnis führen. Das Bundesministerium für Arbeit und Soziales (2017, S. 149) äußert sich in seinem „Weißbuch" zurückhaltend. Darin kündigt es nur an zu prüfen, ob und in welcher Weise in diesem Zusammenhang gesetzgeberischer Handlungsbedarf besteht.

Soweit der Einsatz einer technischen Einrichtung (auch) dem Gesundheitsschutz der Beschäftigten dient – wie dies bei bei Verwendung so genannter Exoskelette der Fall sein kann –, hat der Betriebsrat darüber hinaus ein Mitbestimmungsrecht nach § 87 Abs. 1 Nr. 7 BetrVG (Martini & Botta 2018, S. 634).

7 Fazit und Ausblick

Auch wenn im Rahmen dieser Übersicht viele Themen nur angerissen werden konnten, zeigt sich doch, dass das Verhältnis zwischen Industrie 4.0 und Recht von zahlreichen Wechselwirkungen geprägt ist. Industrie 4.0 und die damit zusammenhängenden technischen Innovationen rufen in der produzierenden Industrie eine Vielzahl von Rechtsfragen hervor. Es wäre jedoch übertrieben zu behaupten, dass die technische Revolution zu einer Revolution des Rechts führen würde. Nicht alle aufgeworfenen Rechtsfragen sind neu. Viele Fragen können bereits mit dem geltenden Recht zufriedenstellend beantwortet werden. In zahlreichen Bereichen genügt es, privatautonome Regelungen mit Hilfe angemessener Vertragsklauseln zu gestalten. In anderen Bereichen besteht jedoch noch gesetzgeberischer Anpassungsbedarf. Zu denken ist hierbei etwa an klarstellende Regelungen zur Inhaltskontrolle von AGB bei vertraglichen Vereinbarungen zwischen Unternehmen, an eine Einführung der Gefährdungshaftung bei vollständig autonomen und selbstlernenden Systemen, an eine Öffnung des Arbeitszeitgesetzes für flexibleres Arbeiten sowie auch an eine Anpassung des Betriebsverfassungsrechts an die Erfordernisse der Digitalisierung. Offene Rechtsfragen müssen zügig geklärt werden, damit das Recht sich nicht als Innovationsbremse für künftige technische und ökonomische Entwicklungen erweist. Vor allem kleinere und mittlere Unternehmen werden vor den technischen Innovationen der Industrie 4.0 zurückscheuen, wenn damit unkalkulierbare rechtliche Risiken verbunden sind. Die weitere technische Entwicklung sollte aufmerksam beobachtet und analysiert werden, um rechtzeitig rechtlichen Regelungsbedarf festzustellen und darauf zu reagieren. Es ist ein Gebot der Stunde, Rechtssicherheit bei der Entwicklung, Einführung und Anwendung von Industrie 4.0-Technologien zu schaffen. Nur wenn es gelingt, neben den technischen Lösungen auch die rechtlichen – aber auch die zahlreichen wirtschaftlichen, sozialen und ethischen – Fragen zu klären und damit die notwendige Akzeptanz bei den Unternehmen und den Beschäftigten zu erreichen, wird die weitere Entwicklung von Industrie 4.0 erfolgreich sein.

Literatur

Baum, F. v., Appt, S. & Schenk, I. (2017a): Die vernetzte Fabrik: Rechtliche Herausforderungen in der Industrie 4.0 (Teil 1). Der Betrieb, 32, S. 1824–1832.

Baum, F. v., Appt, S. & Schenk, I. (2017b): Die vernetzte Fabrik: Rechtliche Herausforderungen in der Industrie 4.0 (Teil 2). Der Betrieb, 33, S. 1888–1893.

Beck, S. (2009): Grundlegende Fragen zum rechtlichen Umgang mit der Robotik. Juristische Rundschau, 6, S. 225–230.

Bitkom (2016a): Bundesverband Informationswirtschaft, Telekommunikation und neue Medien e.V.: Rechtliche Aspekte von Industrie 4.0, Leitfaden (Vorabfassung). Stand: 1. April 2016. URL: https://www.bitkom.org/noindex/Publikationen/2016/Leitfaden/Rechtliche-Aspekte-von-Industrie-40/Bitkom-Leitfaden-Rechtliche-Aspekte-von-Industrie-4-0.pdf (Stand: 25.11.2017).

Bitkom (2016b), Bundesverband Informationswirtschaft, Telekommunikation und neue Medien e.V.: Rechtliche Rahmenbedingungen von Industrie 4.0, Stellungnahme. URL: https://www.bitkom.org/NP-Bitkom/Jahresprogramme/Jahresprogramme/2016/Kriesel-Thomas/Bitkom-Position-Rechtliche-Rahmenbedingungen-Industrie40.pdf (Stand: 25.11.2017).

Bräutigam, P. & Klindt, T. (2015): Industrie 4.0, das Internet der Dinge und das Recht. Neue Juristische Wochenschrift, 16, S. 1137–1142.

Bundesministerium für Arbeit und Soziales (2015): Arbeit weiter denken. Arbeiten 4.0. Grünbuch. Berlin.

Bundesministerium für Arbeit und Soziales (2017): Arbeit weiter denken. Arbeiten 4.0. Weißbuch. Diskussionsentwurf. Berlin.

Bundesverband der Deutschen Industrie e.V. & Noerr (2015): Industrie 4.0 – Rechtliche Herausforderungen der Digitalisierung. Ein Beitrag zum politischen Diskurs. URL: https://bdi.eu/media/presse/publikationen/information-und-telekommunikation/201511_Industrie-40_Rechtliche-Herausforderungen-der-Digitalisierung.pdf (Stand: 25.11.2017).

Giese, K. (2017): Das Mitbestimmungsrecht bei „technischen Einrichtungen" nach § 87 Abs. 1 Nr. 6 BetrVG. In: Baker McKenzie (Hrsg.): Arbeitswelt 4.0 (S. 131–141). Berlin: Erich Schmidt.

Günther, J.& Böglmüller, M. (2015): Arbeitsrecht 4.0 – Arbeitsrechtliche Herausforderungen in der vierten industriellen Revolution. Neue Zeitschrift für Arbeitsrecht, 17, S. 1025–1031.

Hanau, H. (2016): Schöne digitale Arbeitswelt? Neue Juristische Wochenschrift, 36, S. 2613–2617.

Heuer-James, J., Chibanguza, K. & Stücker, B. (2018): Industrie 4.0 - vertrags- und haftungsrechtliche Fragestellungen. Betriebs-Berater, 48, S. 2818–2832.

Hieke, R. (2017): Big Data. Zum gesetzlichen Schutz und der rechtlichen Zuordnung von Daten. Zeitschrift zum Innovations- und Technikrecht, 1, S. 10–20.

Hoeren, T. (2013): Dateneigentum. Versuch einer Anwendung von § 303a StGB im Zivilrecht. MultiMedia und Recht, 8, S. 486–491.

Hornung, G. & Hofmann, K. (2017a): Rechtsfragen bei Industrie 4.0: Rahmenbedingungen, Herausforderungen und Lösungsansätze. In: Reinhart, G. (Hrsg.): Handbuch Industrie 4.0: Geschäftsmodelle, Prozesse, Technik (S. 191–212). München: Carl Hanser.

Hornung, G. & Hofmann, K. (2017b): Industrie 4.0 und das Recht: Drei zentrale Herausforderungen. Publikation der Plattform Industrie 4.0. URL: https://www.plattform-i40.de/I40/Redaktion/DE/Downloads/Publikation/wiss-beirat-i40-recht.pdf?__blob=publicationFile\&v=3 (Stand: 25.11.2017).

Kalbfus, M. (2017): Betriebliche Strukturen in der Arbeitswelt 4.0. In: Baker McKenzie (Hrsg.): Arbeitswelt 4.0 (S. 11–25). Berlin: Erich Schmidt.

Köhler, M. (2017): Alternative Beschäftigungsformen am Beispiel Crowdworking. In: Baker McKenzie (Hrsg.): Arbeitswelt 4.0 (S. 61–80). Berlin: Erich Schmidt.

Krause, R. (2016): Digitalisierung der Arbeitswelt – Herausforderungen und Regelungsbedarf. Gutachten B zum 71. Deutschen Juristentag. München: C.H. Beck.

Küttner, W. (2017): Personalbuch 2017. Arbeitsrecht. Lohnsteuerrecht. Sozialversicherungsrecht (24. Auflage). München: C.H.Beck.

Mantz, R. & Spittka, J. (2017): Datenschutz und IT-Sicherheit. In: Sassenberg, T./Faber. T. (Hrsg.): Rechtshandbuch Industrie 4.0 und Internet of Things. Praxisfragen und Perspektiven der digitalen Zukunft (S. 140–182). München: C.H. Beck.

Martini, M. & Botta, J. (2018): Iron Man am Arbeitsplatz? - Exoskelette zwischen Effizienzstreben, Daten- und Gesundheitsschutz. Neue Zeitschrift für Arbeitsrecht, 10, S. 625–637.

Meyer-Michaelis, I., Falter, K. & Schäfer, A. (2016): Rechtliche Rahmenbedingungen von Crowdworking – Chancen und Risiken dieser Möglichkeit von Fremdpersonaleinsatz – Der Betrieb, 43, S. 2543–2546.

Meyerhans, T. (2017): Die Gestaltungsmöglichkeiten des § 3 BetrVG im Hinblick auf die Arbeitswelt 4.0. In: Baker McKenzie (Hrsg.): Arbeitswelt 4.0 (S. 27–34).

Berlin: Erich Schmidt.

Neighbour, K. (2017): Arbeitsrecht – Realität und Herausforderungen. In: Sassenberg, T./Faber. T. (Hrsg.): Rechtshandbuch Industrie 4.0 und Internet of Things. Praxisfragen und Perspektiven der digitalen Zukunft (S. 211–237). München: C.H. Beck.

Niklas, T. & Thurn, L. (2017): Arbeitswelt 4.0 – Big Data im Betrieb. Betriebs-Berater, 27/28, S. 1589–1596.

Paal, B. & Hennemann, M. (2017): Big Data im Recht. Wettbewerbs- und daten(schutz)rechtliche Herausforderungen. Neue Juristische Wochenschrift, 24, S. 1697–1701.

Plattform Industrie 4.0 (2016): Industrie 4.0 – wie das Recht Schritt hält. Ergebnispapier. URL: https://www.bmwi.de/Redaktion/DE/Publikationen/Industrie/industrie-4-0-wie-das-recht-schritt-haelt.pdf?__blob=publicationFile\&v=6 (Stand: 25.11.2017).

Redeker, H. (2017), IT-Recht (6. Auflage), München: C.H. Beck.

Sassenberg, T. & Faber T. (2017), Rechtshandbuch Industrie 4.0 und Internet of Things. Praxisfragen und Perspektiven der digitalen Zukunft. München: C.H. Beck.

Sattler, A. (2017), Schutz von maschinengenerierten Daten. In: Sassenberg, T./Faber. T. (Hrsg.): Rechtshandbuch Industrie 4.0 und Internet of Things. Praxisfragen und Perspektiven der digitalen Zukunft (S. 27–52). München: C.H. Beck.

Schmidl, M. & Tannen, F. (2017): Das neue Bundesdatenschutzgesetz: die wichtigsten Regelungen für die Unternehmenspraxis. Der Betrieb, 29, S. 1633–1640.

Spindler, G. (2016): Digitale Wirtschaft – analoges Recht: Braucht das BGB ein Update? Juristenzeitung, 17, S. 805–816.

Weichert, T. (2017): Die Verarbeitung von Wearable-Sensordaten bei Beschäftigten. Neue Zeitschrift für Arbeitsrecht, 9, S. 565–570.

Wybitul, T. (2017): Der neue Beschäftigtendatenschutz nach § 26 BDSG und Art. 88 DSGVO. Neue Zeitschrift für Arbeitsrecht, 7, S. 413–419.

JOURNAL OF ENGINEERING, MANAGEMENT AND OPERATIONS 1 (2018)

Informationstechnische Standortverkabelung als grundlegende informationstechnische Infrastruktur: Einblick in die internationale Normungsarbeit

Albrecht M. Oehler[a]

[a] *ESB Business School, Reutlingen University, D-72762 Reutlingen*

Abstract

In diesem Beitrag wird dargestellt, wie internationale Normungsarbeit abläuft. Dies erfolgt am Beispiel der Normung von informationstechnischer Standortverkabelung. Dort werden beispielsweise die Netzwerk-Kabelkategorien, wie Kat.5, definiert.

1 Normung national und international

1.1 Normungsarbeit

Normen dienen dazu, dass grenzüberschreitend Produkte eingesetzt und Märkte geöffnet werden können. Sie ermöglichen den internationalen Handel und Kosteneinsparungen.

Normen sind keine Gesetze. Sie sollen die Funktionalität und Interoperabilität gewährleisten und dafür sorgen, dass Produkte unterschiedlicher Hersteller miteinander agieren können. Sobald die Regierung Gesetze erlässt, welche Normen referenzieren, erhalten diese Gesetzescharakter. Dies kann beispielsweise zum Schutze der Anwender dienen, wie es zum Beispiel bei dem Gesetz zur Elektromagnetischen Verträglichkeit (EMV) der Fall ist.

Die Normungsarbeit ist ein demokratischer Prozess, bei welchem international die jeweiligen Länder und national die jeweiligen, von den Fachkreisen benannten, Experten abstimmen. Bei der Normungsarbeit können unterschiedliche Produkte entstehen: zum einen Normen (en.: standards), welche nur mit einer geforderten Mehrheit angenommen und gegebenenfalls mit einer Sperrminorität abgelehnt werden können. Daher muss die Erstellung entsprechender Dokumente immer im Konsens erfolgen. Neben den Normen gibt es Anwendungsregeln (en.: technical reports), welche nur eine einfache Mehrheit benötigen, um angenommen zu werden. Sie beschreiben beispielsweise die Anwendung aktueller Technologien.

Die Entstehung einer Norm sei anhand einer internationalen Norm erklärt. Um eine neue Norm zu starten muss zunächst ein Antrag gestellt werden, der bereits mehrheitlich unterstützt werden muss (en.: *New Work Item Proposal*, NWIP). Ist das Projekt genehmigt, geht es über einen ersten Entwurf (en. straw man oder *Working Draft*) zu einem Komitee-Entwurf (en.: *Committee Draft*, CD), welcher bei internationalen Normen auf Länderbasis abgestimmt wird und noch von dem Unterkomitee verwaltet wird. Dann kommt der Entwurf der internationalen Norm (en.: *Draft International Standard*, DIS), welcher zentral vom Central Office zur Abstimmung an die Länder verschickt wird. Hier muss eine Zweidrittelmehrheit vorhanden sein und eine Einviertel-Sperrminorität kann die Norm ablehnen. Nach Annahme dieses DIS können noch die letzten technischen Änderungen erfolgen. Wird der DIS genehmigt, so wird ein letzter Entwurf (en.: *Final Draft International Standard*) verfasst und zur Abstimmung an die teilnehmenden Länder verteilt. Diese Abstimmung verläuft auf oberster Ebene und erlaubt nur noch ein Ja oder Nein. Bei Annahme wird dieser Schlussentwurf unverändert als internationale Norm (en.: *International Standard*, IS) veröffentlicht.

Bis spätestens zum DIS erscheint in Deutschland die deutsche Übersetzung des entsprechenden Dokuments. Natürlich gibt es auch für die endgültige Internationale Norm eine deutsche Übersetzung.

Bei der Normungsarbeit wird eine besondere Bedeutung den Hilfsverben zugemessen. Es wird unterschieden in Vorschriften, Empfehlungen und Erlaubnisse. Vorschriften verwenden das Verb müssen (en.: shall), Empfehlungen das Verb sollen (en.: should) und Erlaubnisse das Verb dürfen (en.: may). Um normkonform zu sein, müssen zumindest die Vorschriften, also die „shalls", eingehalten werden.

1.2 DKE

In Deutschland wird die Normungsarbeit bezüglich Elektrotechnik und Informationstechnik bei der Deutschen Kommission Elektrotechnik Elektronik Informationstechnik in DIN (Deutsches Institut für Normung) und VDE (Verband der Elektrotechnik Elektronik Informationstechnik e.V.) (DKE) bearbeitet. Dort gibt es Komitees mit Unterkomitees, welche die verschiedenen Themen behandeln und dabei Zuarbeit zu den entsprechenden internationalen Komitees leisten. Ein Komitee mit dieser Funktion wird Spiegelkomitee genannt. Unterkomitees werden mit von den Fachkreisen, zum Beispiel vom ZVEI (Zentralverband der Elektrotechnik und Elektronikindustrie e.V.) oder ZVEH (Zentralverband der Deutschen Elektro- und Informationstechnischen Handwerke), benannten Experten besetzt. Sie wählen für eine Amtsperiode einen Obmann. Das Unterkomitee kann bei Bedarf Arbeitskreise einsetzen, in welchen Experten niederschwellig, d.h. ohne Benennung, aktiv werden können.

Für die Normung der informationstechnischen Standortverkabelung gibt es eine Besonderheit: diese wird in einem sogenannten Gemeinschaftskomitee[1] (GK) bzw.

[1] Die Bezeichnung „Gemeinschaft" rührt daher, dass das GK 715 das Spiegelkomitee zum internationalen Komitee Gemeinschaftskomitee JTC 1/SC 25 ist (s. Abschnitt 1.3)

in Gemeinschaftsunterkomitee (GUK) bearbeitet. Das GK 715 heißt „Verbindung von Einrichtungen der Informationstechnik“ und hat u.a. das Gemeinschaftsunterkomitee 715.3 „Informationstechnische Verkabelung von Gebäudekomplexen“, welches seinerseits derzeit 9 Arbeitskreise hat.

1.3 ISO/IEC

International existieren zwei wesentliche Instanzen, welche für Normung verantwortlich sind. Zum einen die International Organization for Standardization (ISO), welche für die internationale Normung beispielsweise von mechanischen Komponenten, aber auch von Datentechnik zuständig ist. Auf der anderen Seite gibt es die ältere International Electrotechnical Commission (IEC), welche 1906 gegründet wurde und die Normung aller elektrotechnischer Angelegenheiten übernimmt. Als die bei ISO genormte Schreibmaschine elektrisch wurde und sich die ersten Personal Computer abzeichneten, wurde es notwendig, dass ISO und IEC zusammenarbeiten. Daher wurde 1987 ein Gemeinschaftskomitee gegründet, das Joint Technical Committee (JTC) mit der Nummer 1 und dem Titel „Information Technology“.

Unter diesem JTC 1 gibt es Unterkomitees (en.: Sub Committees, SC), welche für die unterschiedlichen Themen verantwortlich sind. Beispielsweise bearbeitet das SC 25 „Interconnection of Information Technology Equipment“ die Verbindung von informationstechnischen Einrichtungen.

Die eigentliche Arbeit erfolgt in Arbeitsgruppen (en.: Working groups). Bei SC 25 gibt es u.a. die WG 3 „Customer Premises Cabling“ zum Thema informationstechnische Standortverkabelung. Dort arbeiten ca. 60 Experten aus über 20 Ländern und fünf Kontinenten unter der Leitung eines Obmanns (en.: convenor). Abbildung 1 zeigt die WG 3 bei ihrem 61.Treffen in Reutlingen im September 2016.

Abbildung 1: 61.Treffen der Arbeitsgruppe ISO/IEC JTC 1/SC 25/WG 3 an der Hochschule Reutlingen

Abstimmungen erfolgen nur auf Länderebene, d.h. innerhalb des SC 25 bzw. innerhalb des übergeordneten JTC 1. Die Arbeit in der Arbeitsgruppe muss im Konsens erfolgen. Dies stellt häufig eine große Herausforderung dar, zumal wenn die jeweiligen Länder Firmeninteressen vertreten. Hier gilt es Entscheidungen zu treffen bzw. Kompromisse zu erarbeiten und dabei die Mehrheitsverhältnisse zu berücksichtigen. In der Vergangenheit gab es zwei global agierende Firmen, welche in Konkurrenz zueinander standen. Daher gab es etliche Länder, die zu den Treffen Experten einer dieser beiden Firmen entsandten. Abbildung 2 zeigt die Zusammensetzung der Treffen mit einer Zuordnung der entsprechenden Firmeninteressen[2].

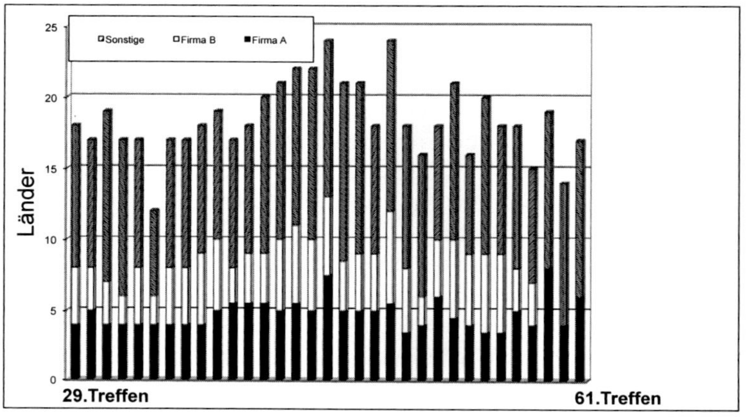

Abbildung 2: Zusammensetzung der Treffen der Arbeitsgruppe ISO/IEC JTC 1/SC 25/WG 3

Ohne die Unterstützung von Experten einer dieser Firmen war es in der Regel nicht möglich nationale Interessen durchzusetzen.

1.4 CEN und CENELEC

Ähnlich wie bei der Normungsarbeit in der ISO/IEC verläuft auch die Normungsarbeit auf europäischer Ebene bei dem European Committee for Standardization (CEN). Für die elektrotechnischen Belange gibt es dort die CENELEC (European Committee for Electrotechnical Standardization). Besonderheit an der europäischen Normungsarbeit ist, dass national nichts genormt werden darf, was bereits auf europäischer Ebene bearbeitet wird. Außerdem gibt es Prozeduren, die eine einfache Übernahme von internationalen Normen als europäische Normen (EN) ermöglichen. Dort muss Augenmerk darauf gelegt werden, dass durch die Übernahme einer internationalen Norm kein Schaden für die nationalen Belange entsteht, wenn die entsprechende europäische Norm anschließend in eine nationale Norm bzw. in nationales Recht umgesetzt wird.

[2] Die Arbeitsgruppe WG 3 trifft sich halbjährlich. Abb. 2 zeigt die Treffen vom 29.Treffen im Juni 2000 in Tromsø, Norwegen, bis zum 61.Treffen im September 2016 in Reutlingen, und die Firmeninteressen nach Einschätzung des Autors.

2 Informationstechnische Standortverkabelung

2.1 Geschichte

Ursprünglich gab es für verschiedene Anwendungen jeweils verschiedene Verkabelungen. Beispielsweise hatte die Firma IBM für ihre Anwendungen die sogenannte Typ 1 Verkabelung. Bei geänderten Anwendungen musste immer neu verkabelt werden.

1995 erschien erstmalig die europäische Norm EN 50173 bzw. die internationale Norm ISO/IEC 11801 für eine anwendungsneutrale strukturierte informationstechnische Standortverkabelung. Diese ermöglicht eine Verkabelung ohne Kenntnis der künftig einzusetzenden Anwendungen und liefert somit einen hohen Investitionsschutz. Sie war ein Bestseller unter den Normen.

2.2 EN 50173 und ISO/IEC 11801

Die Verkabelung für Bürostandorte wird in drei Bereiche eingeteilt. Die Verkabelung beginnt am Standortverteiler (SV) und verläuft innerhalb des Primärbereiches oder Campusbereiches zu den Gebäudeverteilern (GV) der einzelnen Gebäude. Dort beginnt der Sekundärbereich oder Steigebereich, in welchem sternförmig vom Gebäudeverteiler zu den jeweiligen Etagenverteilern (EV) verkabelt wird. Schließlich erfolgt im Tertiärbereich oder Horizontalbereich eine sternförmige Verkabelung zum Teilnehmeranschluss (TA). Diese Struktur ist in Abbildung 3 zu sehen. Sie heißt Baumtopologie, bzw. im Englischen hierarchische Sterntopologie. Der Vorteil dieser Topologie liegt darin, dass gezielt einzelne Stränge ab- oder zugeschaltet werden können, ohne den Betrieb der anderen zu stören. Dies ist im Fehlerfall oder bei sicherheitsbedingten Änderungen von Zugriffsrechten wichtig.

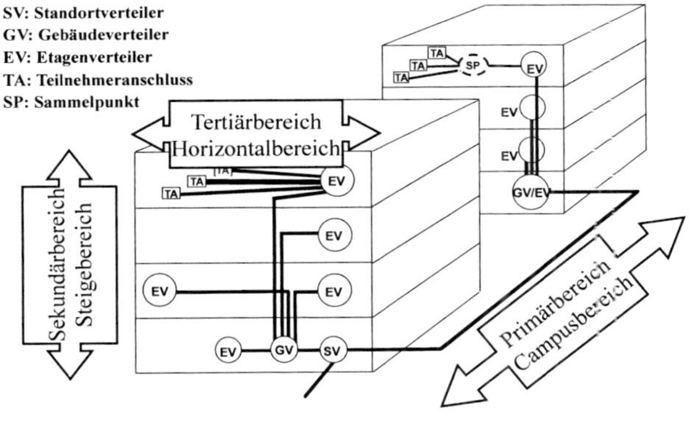

Abbildung 3: Verkabelungsbereiche

Die Verkabelung erfolgt in den Primär- und Sekundärbereichen in der Regel mit Lichtwellenleitern. Dort sind maximale Entfernungen von 2000 m zwischen Standortverteiler und Teilnehmeranschluss, bzw. von 500 m im Steigebereich zu überbrücken.

Im Tertiärbereich, der eine maximale Ausdehnung von 100 m hat, erfolgt die Verkabelung in der Regel mit symmetrischen Kupferkabeln.

Abbildung 4 zeigt die Übertragungsstrecke (en.: channel). Die Übertragungsstrecke geht von einem aktiven Gerät zum anderen, wobei die Anschlüsse an diesen Geräten nicht mehr Teil der Übertragungsstrecke sind.

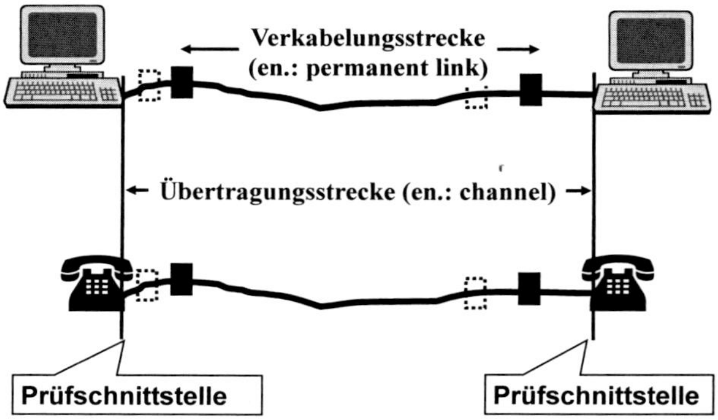

Abbildung 4: Übertragungsstrecke

Die Übertragungsstrecke kann im Horizontalbereich bis zu vier Steckverbindungen beinhalten (s. Abb. 5).

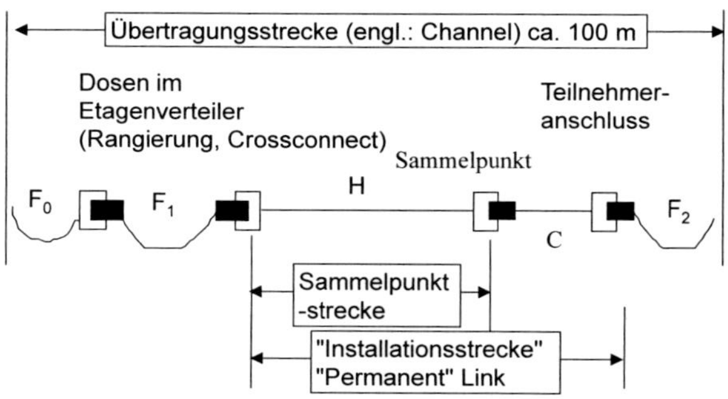

Abbildung 5: Beispielausführung mit vier Steckverbindungen

Am Teilnehmeranschluss, also an der Dose am Arbeitsplatz, ist ein spezielles Steckgesicht vorgeschrieben. Dieses ist im Bürobereich für Kupferverkabelungen der 8-Kontaktstifte Steckverbinder nach IEC 603-7. Er wird im Volksmund „RJ-45" bzw. „Western-plug" genannt.

Spannend ist, dass dieser Steckverbinder ursprünglich für Anwendungen bis 3 MHz entwickelt wurde und Varianten von ihm jetzt bis 2000 MHz spezifiziert werden. Falls auf die Rückwärtskompatibilität verzichtet wird darf auch der IEC 61076-3-104 eingesetzt werden, der schon von Anfang an für höhere Frequenzen entworfen wurde und u.a. ein paarweises Herausführen der einzelnen Paare erlaubt.

Abbildung 6: Steckverbinder IEC 603-7

Abbildung 7: Steckverbinder nach IEC 61076-3-104

Für Lichtwellenleiterverkabelung ist am Teilnehmeranschluss der Duplex-LC-Steckverbinder nach IEC 874-14 vorgeschrieben.

Abbildung 8: Steckverbinder nach IEC 874-14

An allen anderen Orten können beliebige Steckgesichter verwendet werden, solange die physischen und übertragungstechnischen Anforderungen erfüllt werden.

Zwischen dem Teilnehmeranschluss und den Etagenverteiler verläuft die Installationsstrecke (siehe Abbildung 4).

Für die Übertragungsstrecken (vgl. Abb 4) legt die Norm die übertragungstechnischen Eigenschaften fest. Die übertragungstechnischen Eigenschaften der Übertragungsstrecke werden mittels Klassen klassifiziert. Es gibt die Klassen A bis F_A, wobei die Klasse A bis 100 kHz und die Klasse F_A bis 1000 MHz spezifiziert ist. Für Anwendungen in Rechenzentren gibt es ferner die Klassen I und II, welche bis 2000 MHz spezifiziert sind. Die Klassen sind jeweils rückwärtskompatibel. Dies sei am Beispiel der Dämpfung und des Nebensprechens gezeigt. Abb. 9 zeigt zwei aktive Komponenten, welche über zwei Doppeladern (en.: Twisted Pairs) kommunizieren. Dabei wird einerseits das Signal gedämpft und gelangt mit weniger Leistung zum entfernten Empfänger (U_{2A}), angedeutet durch einen weniger stark ausgeprägten schraffierten Pfeil. Je geringer diese Dämpfung ausfällt, desto besser. Andererseits gibt es eine Eigenstörung am Sendeort durch das Nebensprechen auf den eigenen Empfänger (U_{1B}), mit einem gepunkteten Pfeil angedeutet. Je stärker dieses Nebensprechen gedämpft wird, desto weniger wird das Empfangssignal gestört.

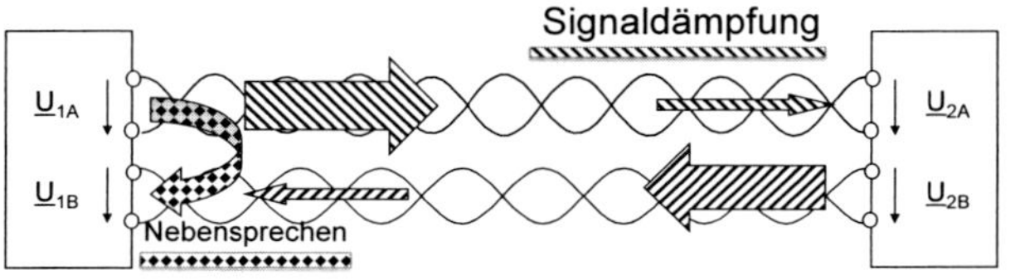

Abbildung 9: Dämpfung und Nebensprechen

Abbildung 10 zeigt die geforderten Werte für die Klassen D, E_A und F_A. Hier wird die Rückwärtskompatibiltät zwischen den Klassen deutlich, d.h. eine höhere Klasse ist besser oder mindestens gleich gut wie alle Klassen unter ihr.

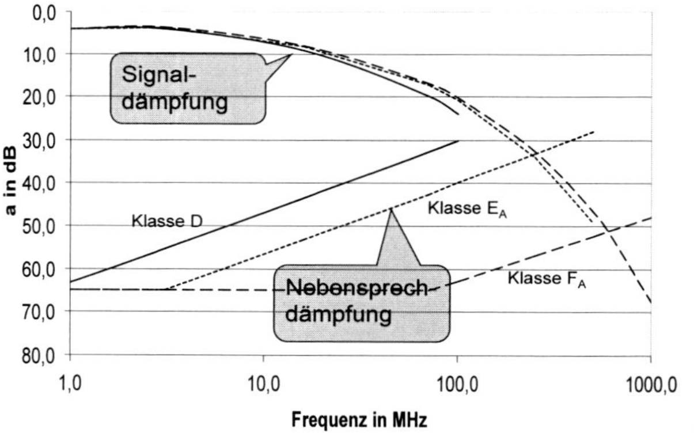

Abbildung 10: Anforderungen an Signaldämpfung und Nebensprechdämpfung

Für Lichtwellenleiterverkabelungen existieren in der internationalen Norm derzeit keine Klassen. Hier muss in Abhängigkeit von der gewünschten Anwendung und dem eingesetzten Lichtwellenleiter die maximale Übertragungsstreckenlänge bestimmt werden. Die Anwendungsneutralität lässt somit zu Wünschen übrig. Daher hat das deutsche Normungskomitee DKE GUK 715.3 Arbeiten für eine deutsche Norm gestartet, um wenigstens auf nationaler Ebene anwendungsneutrale Lichtwellenleiterklassen anbieten zu können.

Voraussetzung für eine normkonforme Verkabelung ist die richtige Wahl des Steckgesichtes am Teilnehmeranschluss und die Einhaltung der Grenzwerte für die Übertragungsstrecke. Eine Möglichkeit diese Werte einzuhalten ist der Beispielausführung (en.: reference implementation) zu folgen. Werden die dort gegebenen Maximallängen nicht überschritten und verwendet man spezielle Komponenten, wird die Übertragungsstrecke die Grenzwerte einhalten.

Unter Verwendung der Beispielausführung (z.B. aus Abb. 5) werden mittels einer Modellierung aus den Übertragungseigenschaften einer Klasse die Anforderungen an Stecker und Kabel abgeleitet. Diese Komponentenanforderungen werden in Kategorien angegeben. Beispielsweise kann die Klasse D mit Komponenten der Kat. 5 realisiert werden, die Klasse F_A mit Komponenten der Kat. 7_A und die Klasse II mit Komponenten der Kat. 8.2.

3 Normungsarbeit

3.1 Konsensbildung

Es ist die Aufgabe des Obmanns eine Gruppe von Expertinnen und Experten so zu lenken, dass ein gemeinsames Dokument im Konsens (!) erstellt werden kann. Im Englischen heißt Obmann „Convenor", es wird also das Zusammenführen der Beteiligten betont. Im Französischen heißt er „Animateur", was die Anregung der Beteiligten hervorhebt. Wann Konsens erreicht wurde entscheidet der Obmann. Der Obmann sollte darauf bedacht sein, möglichst gut die jeweiligen Beweggründe wahrzunehmen und dann einen tragfähigen Kompromiss zu ermitteln. Ein Konsens gilt dann als erzielt, wenn keine anhaltende Opposition mehr besteht. Bei geeigneter Sitzungsführung können auch kontrovers diskutierte Dokumente den Normungsprozess relativ schnell durchlaufen. Dies war zuletzt bei der Schlussabstimmung der aktuellen ISO/IEC 11801-Serie der Fall, die ohne eine einzige Gegenstimme erfolgte.

3.2 Zusammenarbeit mit anderen Gremien

Die Hauptaufgabe der Arbeitsgruppe WG 3 ist die Festlegung von Eigenschaften für Verkabelungssysteme. Damit verfolgt die Arbeitsgruppe den Systemansatz und handelt als Systemkomitee. Hierfür ist eine Kommunikation mit anderen Gremien notwendig. Zunächst müssen die Anforderungen der Anwendungsgremien erfüllt werden, bzw. muss den Anwendungsgremien mitgeteilt werden, welche Eigenschaften die Verkabelung erfüllen kann und wird. Zweitens müssen Randbedingungen wie Elektromagnetische Verträglichkeit berücksichtigt werden. Schließlich muss mit Komitees kommuniziert werden, welche für die Normung von Komponenten wie Kabel und Steckverbinder zuständig sind. Diese müssen darüber informiert werden, welche Anforderungen die Komponenten benötigen, bzw. muss von diesen in Erfahrung gebracht werden, wie weit etwaige Anforderungen gehen können.

Diese Kommunikation erfolgt durch sog. „liaisons", dargestellt durch die Verbindungspfeile in Abb. 11.

Die liaisons sind benannte Experten, die in beiden Komitees aktiv teilnehmen und jeweils berichten. Eine transparente und reibungslose Zusammenarbeit ist essentiell.

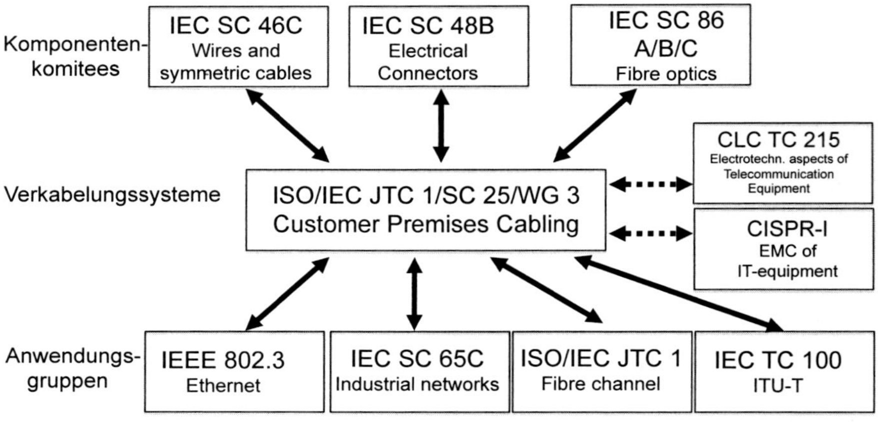

Abbildung 11: Zusammenarbeit mit anderen Gremien

3.3 Kommunikationskultur

Bei der Zusammenarbeit mit Expertinnen und Experten von vielen verschiedenen Ländern sollte berücksichtigt werden, dass diese verschiedene kulturelle Hintergründe haben, somit anders agieren und sich an Diskussionen unterschiedlich beteiligen. Häufig verlaufen die Diskussionen nur zwischen wenigen Experten/innen. Wichtig ist es dann auch die schweigenden Teilnehmenden zu Wort kommen zu lassen. Ein oft ausgetauschtes Argument muss noch lange nicht die Meinung der Mehrheit darstellen.

4 Zusammenfassung

Dieser Aufsatz gibt einen Einblick in die Entstehung von Normen und in den Sprachgebrauch. Als Beispiel wird die Norm für die Informationstechnische Standortverkabelung verwendet. Er erläutert die die Normungsarbeit im Hinblick auf Konsensbildung innerhalb der Arbeitsgruppe und auf die Zusammenarbeit mit anderen Gremien.